Macromolecular Systems – Materials Approach

Springer

Berlin
Heidelberg
New York
Barcelona
Budapest
Hong Kong
London
Milan
Paris
Santa Clara
Singapore
Tokyo

Hans Zweifel

Stabilization
of Polymeric Materials

With 84 Figures

 Springer

Dr. Hans Zweifel
Ciba Specialty Chemicals Inc.
Additives Division, BU AP
Basel, Switzerland

Editors of the professional reference book series
Macromolecular Systems – Materials Approach are:
A. Abe, Tokyo Institute of Polytechnics, Iiyama
L. Monnerie, Ecole Superieure de Physique et de Chimie Industrielles, Paris
V. Shibaev, Moscow State University, Moscow
U. W. Suter, Eidgenössische Technische Hochschule Zürich, Zürich
D. Tirrell, University of Massachusetts at Amherst, Amherst
I. M. Ward, The University of Leeds, Leeds

For this volume the responsible editor is:
U. W. Suter, Eidgenössische Technische Hochschule Zürich, Zürich

ISBN 3-540-61690-X Springer-Verlag Berlin Heidelberg New York

Library of Congress Cataloging-in-Publication Data
Zweifel, Hans, 1939–   . Stabilization of polymeric materials / Hans Zweifel.   p.   cm. – (Macromolecular systems, materials approach). Includes bibliographical references and index.
  ISBN 3-540-61690-X (acid-free paper)
  1. Polymers – Additives. 2. Stabilizing agents. I. Title. II Series.
TP1142.Z94   1997   668.9–dc21   97-42012   CIP

© Springer-Verlag Berlin Heidelberg 1998
Printed in Germany

Coverdesign: de'blik, Berlin
Typesetting: K+V Fotosatz GmbH, Beerfelden

SPIN 10101311   2/3020-5 4 3 2 1 0 – Printed on acid-free paper

Preface

Polymers undergo oxidative degradation reactions in the course of processing and actual use. Depending on molecular structure and conditions of use, material properties can change drastically. For this reason, adequate stabilization against oxidative degradation is mandatory.

This monograph describes the most important principles of oxidative degradation reactions and their inhibition by suitable stabilizers. The scope was intentionally chosen to include oxidative degradation of polymers and its inhibition. Consequently, description of stabilization and inhibition of PVC was omitted, because it involves essentially dehydrochlorination and its delay.

The terms relating to degradation, aging, and related chemical transformations of polymers follow essentially the guidelines as given in International Union of Pure and Applied Chemistry, Macromolecular Division, Commission on Macromolecular Nomenclature, Provisional Recommendations 1994.

The monograph would not have been possible without the help of my colleagues V. Dudler, B. Gilg, F. Gugumus, G. Knobloch, E. Kramer, Ch. Kroehnke, P. Michaelis, J.-R. Pauquet, P. Rota-Graziosi, B. Rotzinger, A. Schmitter, T. Schmutz, C. Sorato, R.V. Todesco and J. Zingg.

Thanks are due to N.C. Billingham and J. Pospisil for numerous discussions and contributions during the preparation of the monograph.

I am particularly grateful to Mrs. C. Guenin for the tremendous workload involved in writing and the preparation of the graphs.

Basel, November 1997   *Hans Zweifel*

Contents

Chapter 2
Principles of Stabilization 41

Chapter 3
Principles of Stabilization of Individual Substrates 71

CHAPTER 1

Principles of Oxidative Degradation

1.1
Introduction

Organic materials undergo degradation reactions in the presence of oxygen. As a result, numerous oxidation products are formed such as, e.g. peroxides, alcohols, ketones, aldehydes, acids, peracids, peresters or γ-lactones. Elevated temperatures, heat, and catalysts, such as metals and metal ions, assist oxidation. Degradation products arising from the oxidation of defined low molecular weight hydrocarbons can be relatively easily isolated and analyzed.

Most of the known polymers have structural elements that, in analogy with low molecular compounds, are prone to oxidative degradation reactions. Because these are macromolecules, isolation of the resulting oxidation products and their identification is difficult.

Spectroscopic investigations show that the degradation products contain the same functional groups as those formed by the oxidation of low molecular weight hydrocarbons.

Thermoplastic polymers are prepared by chain polymerization, polyaddition, or poly-condensation reactions. Their subsequent processing, usually in several processing steps, leads to the end product. The polymer is subjected to heat and mechanical shear in the course of processing. The finished article has to be used over years without changes in chemical, physical, mechanical and aesthetical properties. Oxygen, heat, light and water are always present throughout the whole cycle. Because under these conditions polymer chains are cleaved by oxidative degradation reactions, and chain branching and crosslinking reactions can also occur, understanding of oxidative degradation and its prevention is extremely important for the use of polymeric products.

Furthermore, the morphology of polymers is influenced by their physical aging, e.g. by subsequent crystallisation or relaxation processes which take place below the melting point or the glass transition temperature. Chemical aging of plastics can be influenced by physical aging. In, e.g. polyesters and polyamides, hydrolysis of the ester, and amide bonds also leads to chain scission.

Modern polymeric parts and end products are complex systems generally containing, along with the polymer, fillers, pigments, reinforcing materials and additives of various types that can also influence oxidative degradation reactions.

This monograph deals exclusively with the oxidative degradation of thermoplastic polymers and its prevention by the addition of suitable stabilizers and stabilizer systems. The behavior of plastics under thermomechanical (processing of melts), thermooxidative (long term stability under thermal stress) and photooxidative (effect of light) conditions is discussed.

1.2
Autoxidation

The phenomenon of oxidation of polymers was investigated very early on in connection with the aging of natural rubber. Hoffmann [1] realised the connection between aging and the absorption of oxygen. Based on the fact that hydrocarbon compounds react with molecular oxygen forming oxidation products, the scheme of autoxidation was developed depicting it as a free radical-initiated chain reaction. Bolland and Gee [2, 3] have shown that the oxidation of hydrocarbons proceeds autocatalytically. The reaction is slow at the start, generally associated with a short induction period and accelerates with the concentration of the resulting hydroperoxides. The free radical-initiated chain reaction of autoxidation, as other radical reactions, can be regarded as proceeding in three distinct steps: chain initiation, chain propagation, and chain termination (Scheme 1.1).

Autoxidation is depicted schematically in Fig. 1.1.

The origin of the primary alkyl radical $R^{\bullet}$ (Eq. 1.1) as initiating species for the chain reaction is still controversial. Direct reaction of hydrocarbons with molecular oxygen in a bimolecular reaction is not favoured because of thermodynamic and kinetic considerations. One explanation is that in the course of polymerisation, adventitious catalysts such as transition metals, radical initiators, impurities in the monomers, and minute amounts of oxygen react and form peroxy radicals $ROO^{\bullet}$ (Eq. 1.2) which abstract hydrogen from the polymer and form an alkyl radical (Eq. 1.3). The technical preparation of "purest" polymer is simply not possible. Structural defects and impurities cannot be excluded. Furthermore, during the first processing step of the melt, e.g. extrusion, blow molding, or injection molding, additional peroxide radicals are formed by the reaction with molecular oxygen under the effect of heat and mechanical shear. Subsequently, they

Chain initiation:

$$\left.\begin{array}{l} R-H \\ R-R \end{array}\right\} \longrightarrow R\cdot \qquad (1.1)$$

Chain propagation:

$$R\cdot + O_2 \longrightarrow ROO\cdot \qquad (1.2)$$

$$ROO\cdot + RH \longrightarrow ROOH + R\cdot \qquad (1.3)$$

$$RO\cdot + RH \longrightarrow ROH + R\cdot \qquad (1.4)$$

$$\cdot OH + RH \longrightarrow H_2O + R\cdot \qquad (1.5)$$

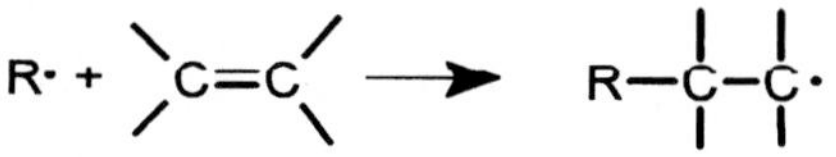

$$R\cdot + ^{\diagdown}_{\diagup}C=C^{\diagup}_{\diagdown} \longrightarrow R-\overset{|}{\underset{|}{C}}-\overset{|}{\underset{|}{C}}\cdot \qquad (1.6)$$

$$R^1-\overset{O\cdot}{\underset{\underset{R^3}{|}}{\overset{|}{C}}}-R^2 \underset{\text{β-Scission}}{\longrightarrow} R^1-\overset{O}{\overset{\|}{C}}-R^2 + \cdot R^3 \qquad (1.7)$$

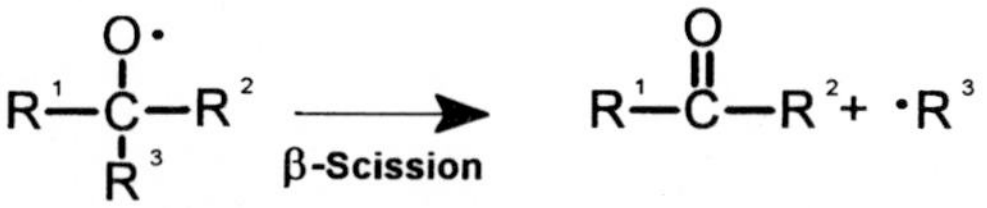

$$R\cdot \xrightarrow{\text{Fragmentation}} \text{Olefin} + R'\cdot \qquad (1.8)$$

Chain branching:

$$ROOH \longrightarrow RO\cdot + \cdot OH \qquad (1.9)$$

$$2ROOH \longrightarrow RO\cdot + ROO\cdot + H_2O \qquad (1.10)$$

Chain termination:

$$R\cdot + ROO\cdot \longrightarrow ROOR \qquad (1.11)$$

$$R\cdot + R\cdot \longrightarrow R-R \qquad (1.12)$$

$$R\cdot + RO\cdot \longrightarrow R-O-R \qquad (1.13)$$

$$2\,R\cdot \xrightarrow{\text{Disproportionation}} RH + \text{Olefin} \qquad (1.14)$$

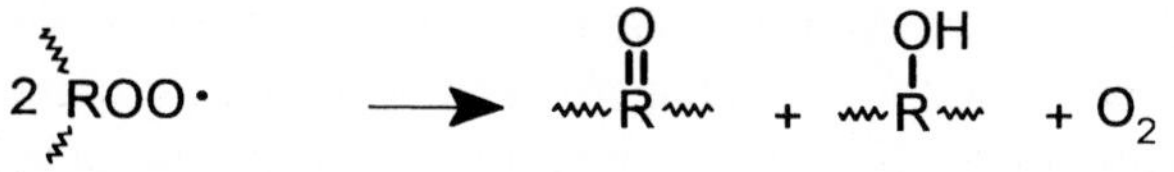

$$2\,\overset{\diagup}{\underset{\diagdown}{R}}OO\cdot \longrightarrow \text{\sim\sim}R\overset{O}{\overset{\|}{\text{\sim\sim}}} + \text{\sim\sim}R\overset{OH}{\underset{|}{\text{\sim\sim}}} + O_2 \qquad (1.15)$$

Scheme 1.1. General diagram of autoxidation. R stands for the polymer backbone. (See Appendix 1)

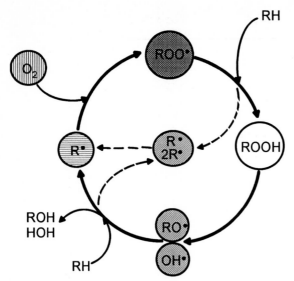

Fig. 1.1. The cycle of autoxidation

$$M^{n+} + ROOH \longrightarrow M^{(n+1)} + RO^{\bullet} + HO^{-} \qquad (1.16)$$

$$M^{(n+1)} + ROOH \longrightarrow M^{n+} + ROO^{\bullet} + H^{+} \qquad (1.17)$$

Scheme 1.2. Decomposition of hydroperoxides initiated by metal ions

$$|\overline{O}=\overline{O}| \equiv |\overline{O}-\overline{O}|$$

Scheme 1.3. Molecular oxygen

form hydroperoxides upon abstraction of hydrogen (Eq. 1.3). The rate of decomposition of hydroperoxides (Eq. 1.9) to alkoxy, and hydroxyl radicals increases with rising temperature. It can proceed, however, under the influence of light or metal ions (Eqs. 1.16 and 1.17, Scheme 1.2).

Alkyl radicals react with molecular oxygen practically without activation energy forming peroxy-radicals. Molecular oxygen is a "diradical" in a triplet state with two electrons in separate, parallel antibonding orbitals (Scheme 1.3).

The rate constant for the reaction of most alkyl radicals with oxygen is [4] of the order of $10^{7}–10^{9}$ mol^{-1}s^{-1}. The abstraction of a hydrogen, H, by a peroxy radical ROO$^{•}$ (Eq. 1.3) requires the breaking of a C-H bond. This reaction (Eq. 1.3) requires a corresponding activation energy and is, therefore, the rate determining step in autoxidation. The rate of the abstraction reaction decreases in the following order: hydrogen in a-position to a C=C double bond ("allyl") > benzyl hydrogen and tertiary hydrogen > secondary hydrogen > primary hydrogen [5–10].

Primary and secondary peroxy radicals are more reactive than the analogous tertiary radicals [6, 7] with regard to hydrogen abstraction. The most reactive are acylperoxy radicals [11].

If sufficient oxygen is available and the formation of peroxy radicals does not occur at too high a temperature, then chain termination mainly proceeds according to Eq. (1.15) by recombination of peroxy radicals [12]. Under oxygen deficient conditions, i.e. if the concentration of [R$^{•}$] is much higher than that of [ROO$^{•}$], chain termination is caused by recombination with other available radical species according to Eqs. (1.11) and (1.12). An important termination reaction is disproportionation of alkyl radicals according to Eq. (1.14).

Most of the above-mentioned findings are based on studies of the oxidation of low molecular weight hydrocarbon compounds. The oxidation of polyolefins and other plastics with hydrocarbon structures proceeds similarly.

Because this is a reaction between a solid phase, the plastic material, or its high viscosity melt and a gas – oxygen, all reactions are diffusion dependent. Knowledge of oxygen's diffusion and solubility in plastics is a prerequisite for understanding the complex oxidation processes in the polymeric matrix.

1.3
Diffusion and Solubility of Oxygen in Polymers

The permeability of various polymers to gases such as oxygen can be relatively easily measured. The determination of its components – diffusion and solubility is much more difficult and the values for only a few polymers are known (Table 1.1).

Billingham [19] drew the following conclusions regarding the role of diffusion and solubility of oxygen in the oxidation of polymers:

(i) the solubility of oxygen in solid polyolefins is only slightly lower than in liquid hydrocarbons
(ii) the diffusion of oxygen, however, is two orders of magnitude slower

Table 1.1. Diffusion coefficients and solubilities of oxygen in some polymers

Polymer	Diffusion Coefficient	Solubility	Temp.	Ref.
	10^7 (cm²/s)	10^3 (mol/kg)	°C	
Hexane	300	~ 5	25	[13, 19]
Natural Rubber	16	5.0	25	[14]
PE-LD	5.4	0.44	25	[15]
PE-HD	1.6	0.68	25	[15]
PMMA	0.12	5.0	25	[16]
PC	0.56	7.43	35	[17]
PA 6.6	0.12	0.61	25	[18]

(iii) the diffusion of oxygen depends on the morphology of the polymer: oxygen is, for example, dissolved exclusively in the amorphous phase of polyolefins.

The mechanical properties of polymers, e.g. polyolefins, are determined to a large extent by the entanglement of the polymer molecules in the amorphous regions. For this reason, oxidative degradation only in these regions leads to loss of strength in such materials.

Generally, and at moderate temperatures, the diffusion of oxygen leads to an equilibrium, regarding oxygen saturation. The concentration of dissolved oxygen is sufficient to transform the alkyl radicals into their corresponding peroxy radicals, i.e. $[ROO^\bullet] > [R^\bullet]$. Diffusion-limited oxidation effects occur when oxygen consumption in polymers exceeds the rate at which oxygen can be supplied by diffusion to the polymer, i.e. $[R^\bullet] > [ROO^\bullet]$. Both "limiting situations" occur in polymers in the course of their life cycle: oxygen-deficient conditions during processing in the melt, and, generally, oxygen saturation in the course of the end product's use.

Unstabilized polypropylene is rapidly degraded by the effect of heat, mechanical shear and radiation with UV light of 300 nm [20]. Chain scission causes drastic reduction of the average molecular weight, M_w. The resulting oxidation products absorb in the infrared region, among others, at 1720 cm^{-1}. Figure 1.2 shows the relationship between the amount of oxidation products and the corresponding average molecular weight, M_w at different exposures of the unstabilized polypropylene (multiple extrusion, oven aging, light exposure).

In oxygen-deficient conditions, as in an extruder, the amount of oxidation products after fivefold extrusion at 280 °C is approximately 30–40 times lower than that after 7 h aging in an air circulating oven at 135 °C or after 250 h light exposure in a XENO-1200 exposure de-

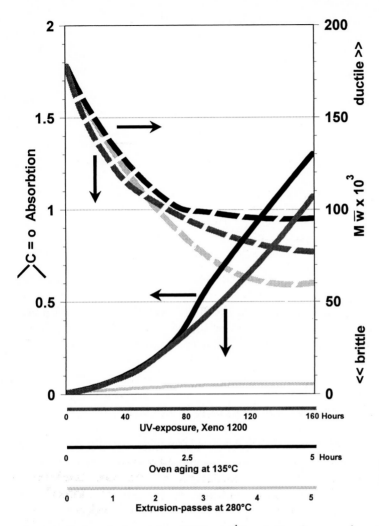

Fig. 1.2. Carbonyl absorption (1720 cm^{-1}) and M_w changes after multiple extrusion passes at 280 °C, oven aging at 135 °C and exposure to UV-irradiation (XENO 1200, b.p. temp. 55 °C)[1]

---

[1] Abbreviations – see Appendix 1

Table 1.2. Factors governing polymer's degradation

Factors Governing The Polymer's Degradation	Polymer Processing [e.g. Extrusion]	Polymer's Service Lifetime [e.g. Heat, Irradiation]
Physical State of the Polymer	Melt	Solid State
Temperature	180°C - 320 °C	-30°C - 150°C
Oxygen Content	Oxygen Deficient Conditions	Generally Oxygen in Saturation Equilibrium
Exposure Time	Minutes	Hours to Years
R$^{\bullet}$ -Concentration	[R$^{\bullet}$] >> [ROO$^{\bullet}$]	[ROO$^{\bullet}$] >> [R$^{\bullet}$]
ROOH -Concentration	Low	High
Rate of ROOH -Decomposition	Fast	Moderate, according to Temperature

vice at 55 °C. The reason is that in the latter two procedures, diffusing oxygen ensures oxygen saturation.

The following chapters deal with the thermooxidative and photooxidative processes in various polymers under a variety of conditions such as processing of the melt, thermal stress and the effect of light (Table 1.2).

1.4
Degradation of Polymers During Processing in the Melt

Following their production in the reactor, polymers usually precipitate as powders or crumbs. Up to the manufacture of the end product, polymers are subjected to one or several processing steps in the melt, e.g. in extrusion, film or blow molding or injection molding. The polymer melt is exposed to high shearing forces in the extruder (Fig. 1.3).

The high mechanical forces, particularly in the region of entangled polymer molecules, lead to C-C chain scission resulting in macroalkyl radicals, R$^{\bullet}$, and simultaneous reduction of the molecular weight M_w [21]. Degradation caused by shear is referred to as thermo-mechanical degradation (Eq. 1.18, Scheme 1.4).

The oxygen dissolved in the polymer reacts with the alkyl radicals forming peroxy radicals, ROO$^{\bullet}$, and subsequently, after abstraction of hydrogen (Eqs. 1.2 and 1.3), leading to hydroperoxides and new alkyl radicals. These hydroperoxides decompose rapidly to the corresponding alkoxy, RO$^{\bullet}$, and hydroxy radicals, $^{\bullet}$OH, at the usual processing temperatures of approximately 180 to over 300 °C. The latter can form

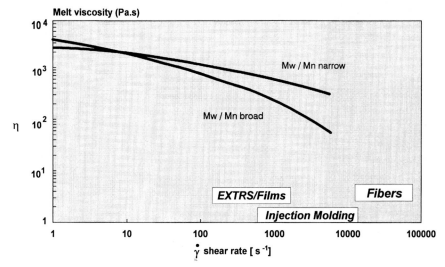

Fig. 1.3. Relation between shear rate and melt viscosity and the influence of molecular weight distribution in typical processing conditions

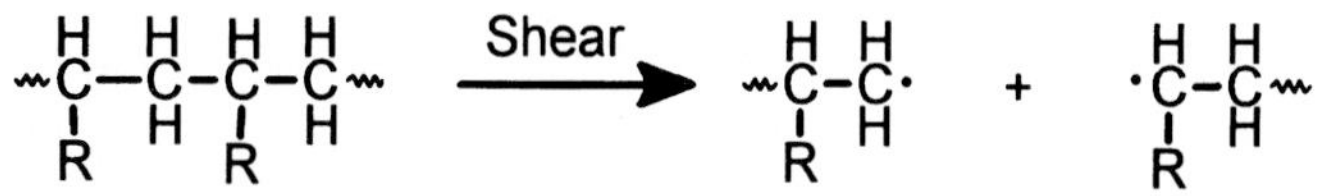

$$\underset{\substack{\text{www}}}{\overset{\substack{H \quad H \quad H \quad H}}{C-C-C-C}}_{\substack{R \quad H \quad R \quad H}} \xrightarrow{\text{Shear}} \underset{\substack{\text{www}}}{\overset{\substack{H \quad H}}{C-C}}^{\bullet}_{\substack{R \quad H}} + \overset{\bullet}{\underset{\substack{\text{www}}}{\overset{\substack{H \quad H}}{C-C}}}_{\substack{R \quad H}} \quad (1.18)$$

Scheme 1.4. Alkyl radical formation upon polymer processing

the inactive products ROH and H_2O and further alkyl radicals, $R^{\bullet}$ (Eqs. 1.4 and 1.5) through hydrogen abstraction. On the other hand, β-scission leads to chain scission of the macromolecule.

In the course of processing in the machine, the oxygen dissolved in the polymer melt is consumed and follow-up diffusion is not possible. It follows that the concentration of alkyl radicals $[R^{\bullet}]$ is much greater than that of peroxy radicals $[ROO^{\bullet}]$.

The chain reaction of autoxidation has, therefore, to proceed through the available alkyl radicals. Possible reactions are shown in Scheme 1.5.

Fragmentation of a macroalkyl radical leads to a decrease of M_W (Eq. 1.8) whereas the disproportionation reaction of two macroalkyl radicals takes place without changes of the macromolecule with regard to M_w (Eq. 1.14). Recombination of two macroalkyl radicals leads to an increase of M_W (Eq. 1.12). The addition of alkyl radicals to -C=C-double bonds (Eq. 1.6) depends to a large extent on steric factors. Yachigo et al. [22] and Knobloch [23] have shown that the processing

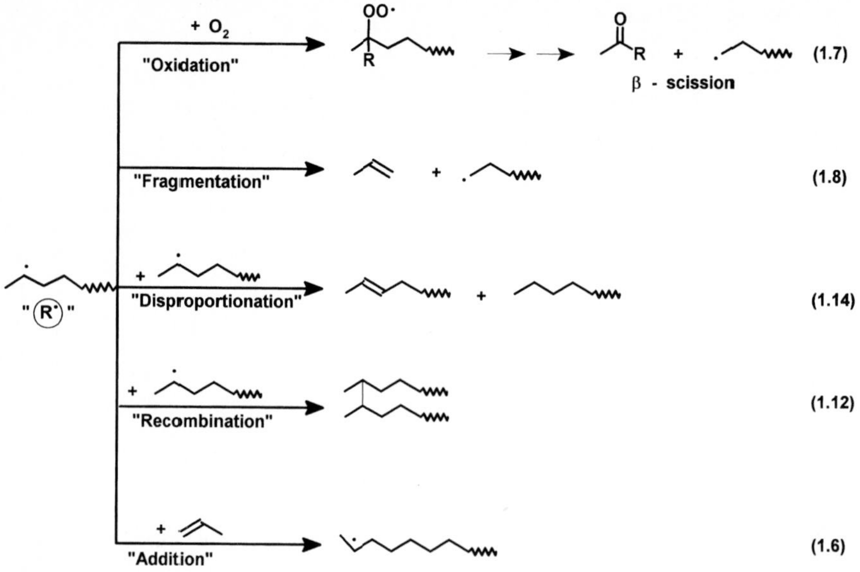

Scheme 1.5 Termination reactions involving alkyl radicals

of a styrene-isoprene-styrene block copolymer in a Brabender kneader under inert atmosphere results in molecular weight reduction. Yet, analogous processing of a styrene-butadiene-styrene block copolymer leads to gel formation caused by crosslinking (Scheme 1.6).

Processing of polypropylene and polystyrene leads in general to a degradation of the polymer caused by chain scission. Processing of polyethylene leads to degradation as well as to chain branching and crosslinking. The polymer's behavior depends on processing conditions such as temperature and shear and also on the structure of the macromolecules (e.g. number of terminal vinyl groups) [24, 25]. Figure 1.4 shows the changes of melt viscosity, η_0, of an unstabilized polypropylene measured at 240 °C and of an unstabilized polyethylene (Cr-catalyst), measured at 221 °C, after a number of extrusions at 260 °C.

1.4.1
Polypropylene

Polymers with branched alkanes as repeating units react with peroxides by intramolecular hydrogen abstraction particularly easily, forming hydroperoxide sequences if a six membered transition state is possible (Scheme 1.7, Eq. 1.19).

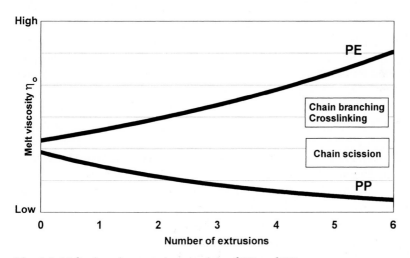

Scheme 1.6. Influence of side chain substituent on chain scission or crosslinking during processing

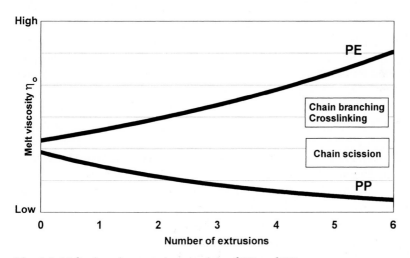

Fig. 1.4. Melt viscosity, upon processing of PP and PE

Polypropylene undergoes pronounced molecular weight degradation in the course of processing. Table 1.3 lists the molecular weights, M_w,

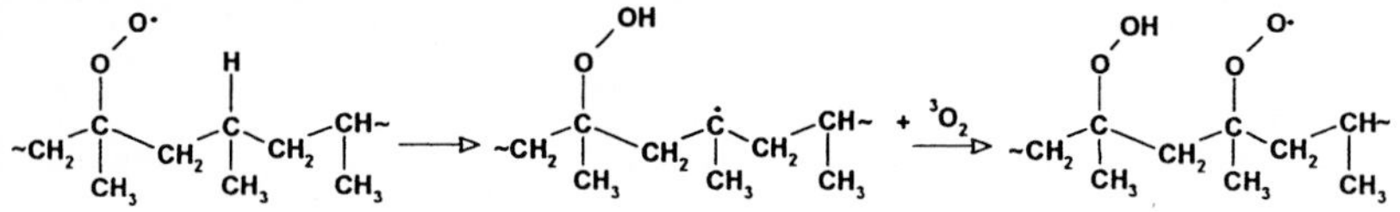

Scheme 1.7. Reaction of molecular oxygen with a polypropylene macroalkylradical

Table 1.3. Change in M_w and M_w/M_n after multiple extrusion of polypropylene at 260 °C

Extrusion Pass #	M_w	M_w / M_n
Virgin PP	270,000	2.99
1	178,000	2.48
3	72,000	1.92
5	60,000	1.71

and the molecular weight distribution, M_w/M_n of polypropylene as function of multiple extrusions at 260 °C.

The concentration of resulting oxidation products is low. IR spectroscopic investigations revealed the formation of carboxylic acid, peracid, ester, perester, carbonyl and γ-lactone groups [20]. Thermo mechanical degradation contributes to a reduction of the polydispersity M_w/M_n. In any event, the contribution of thermo mechanical degradation depends on temperature and shear conditions in the extruder [26].

1.4.2
Polyethylene

Ethylene is polymerized with the help of a variety of catalysts. The resulting polymers differ with regard to density, branching, and the number of terminal vinylic double bonds. Some physical/chemical data for two different PE-HD types are summarized in Table 1.4 [27]. During processing, polyethylene, and particularly high density polyethylene, undergoes degradation by chain scission as well as chain branching, and in extreme cases also crosslinking. Similar reactions are observed also during processing of PE-LLD [24].

Recently "hybrid catalysts" have been frequently used, thus giving rise to polyethylene types having the characteristics of the Cr-type as well as those of the Ti-type.

Table 1.4. Physical/chemical data for unstabilized Cr- and Ti-catalyst type PE-HD

PE-HD	M_W	T_m °C	Density g/cm^3	Unsaturation per 10^6 C Atoms		
				Vinyl Groups	Transvinylene Groups	Vinylidene Groups
Cr-Catalyst type	99,000	131	0.952	993		
Ti-Catalyst type	129,000	129	0.944	115	10	33

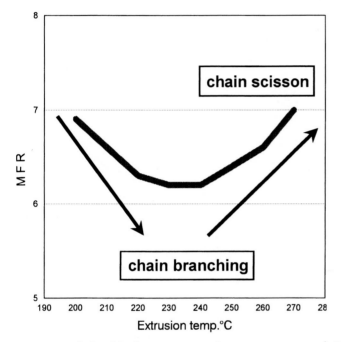

Fig. 1.5. Relationship between extrusion temperature and MFR of the extruded product

Structural factors influence polyethylene's degradation, however, processing conditions also playing a decisive role in whether chain degradation or chain branching (and crosslinking) are the dominant reactions. Extrusion of a Cr-type PE-HD at different temperatures leads to products with different melt mass-flow rate, MFR[1]. The dependence of MFR on extrusion temperature is depicted in Fig. 1.5.

---

[1] Melt mass-flow rate – see Appendix 2

$$\underset{H}{\overset{O\cdot}{\underset{|}{\overset{|}{C}}}}-CH_2-CH_2\sim \quad \xrightarrow{\beta \text{ - Scission}} \quad \sim \overset{O}{\overset{||}{CH}} \; + \; \cdot CH_2-CH_2\sim \qquad (1.7)$$

Scheme 1.8. β-Scission of an alkoxy radical in polyethylene

The diagram shows that at extrusion temperatures in the range of 200–230 °C chain branching dominates, while at higher extrusion temperatures chain scission is the dominating reaction. Campbell and coworkers [28] the processing of a PE-HD in a corotating intermeshing twin screw extruder and have shown that the chosen process parameters determine whether degradation or crosslinking takes place.

Chain branching and crosslinking reactions are caused by the addition of alkyl radicals to vinylic -C=C- double bonds (Eq. 1.6) [27], analogous to the degradation of styrene-butadiene-styrene block copolymers [23, 24]. Degradation by chain scission proceeds to a large extent by β-scission of an alkoxy radical (Scheme 1.8) according to Eq. (1.7) and by thermo mechanical degradation (Eq. 1.18).

1.4.3
Polystyrene

The behavior of polystyrene under processing conditions is similar to that of polypropylene. The accumulation of hydroperoxides becomes possible by means of intramolecular hydrogen abstraction through a six-membered transition state.

Table 1.5 summarizes [29] some results regarding the behavior of crystal polystyrene in the course of multiple extrusions at 190 °C.

The thermooxidative behavior of polystyrene is influenced by structural differences in the backbone such as head-to-head or head-to-tail bonding [30] or so-called "weak links", e.g. R-O-O-R groups, postulated because of polystyrene's production by means of radical initiators [31].

Table 1.5. Change in M_w and M_w/M_n after multiple extrusion of polystyrene at 190 °C

Extrusion Pass #	M_w	M_w / M_n
Virgin PS	335,000	3.2
7	210,000	2.7

1.4.4
Polyamides and Polyesters

Aliphatic polyamides and polyesters react preferentially at the methylene group vicinal to N- or O- (ester or amide bond) by hydrogen abstraction and subsequent reaction with molecular oxygen, forming peroxy radicals. Aliphatic polyamides and also the corresponding polyesters are quite stable under oxygen-deficient processing conditions with regard to molecular weight. However, both substrates undergo molecular weight degradation in the presence of water as a result of hydrolysis. It follows that complete removal of humidity is mandatory prior to any processing. Furthermore, discoloration may occur in unstabilized PA and PET polymers upon processing.

1.4.5
Polyacetals

The only polyacetals of commercial importance are prepared by cationic or anionic polymerisation of formaldehyde or trioxane. Because of the relatively low ceiling temperature, deformylation takes place. For this reason such polyacetals are thermally unstable and have to be transformed into a thermodynamically stable polymer by suitable terminal groups ("end capping"). Frequently copolymers are produced

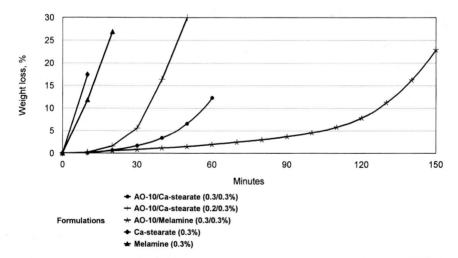

Fig. 1.6. Influence of costabilizers upon thermal treatment of polyacetal copolymer plaques (1 mm), thermogravimetry, 220 °C, air, isothermal. For stabilizer structures see Appendix 3

using ethylene oxide, cyclic acetals or lactones. Deformylation forming formaldehyde starts at approx. 160 °C [32] in a "zipper-like" depolymerization reaction. This reaction stops in copolymers at the first -C-C- bond. The deformylation reaction is acid-catalyzed, hence the use of bases and antioxidants as stabilizers is mandatory in the processing of polyacetals as shown in Fig. 1.6.

1.4.6
Polycarbonate

Prior to processing, polycarbonate has to be carefully dried. Basic or acidic impurities in the substrate lead to chain degradation during processing.

1.4.7
Polyurethanes

Polyurethanes are prepared by addition of polyols (polyether polyols and polyester polyols) to multifunctional isocyanates. Polyether polyols are particularly oxygen-sensitive. Thermooxidative degradation occurs predominantly in the production of flexible foams when water is used for the generation of carbon dioxide as "blowing agent". The manufacture of large polyurethane blocks is exothermic and the resulting temperatures can exceed 160 °C in the interior of such blocks. Figure 1.7 shows the development of temperature in the interior of a polyurethane block (2×1×5 m, density: 21 kg/m^3).

Because of thermooxidative processes a yellow/brown discoloration occurs in the presence of oxygen ("scorch") and the danger of spontaneous ignition is high at temperatures over 170 °C. Poly (ether polyols) used for the production of flexible foams, therefore, always contain antioxidants to prevent uncontrolled thermal autoxidation. Polyurethanes with poly (ester polyol) segments are less sensitive to thermooxidative degradation than poly (ether polyol), although they can undergo degradation reactions as a consequence of hydrolysis.

1.4.8
High Performance Engineering Thermoplastics

The high temperature resistant thermoplastics ("HT-thermoplastics) include polysulfones, poly (ether sulfones), poly (phenylene sulfides), poly (ether imides), poly (aryl-ether-ketones), and poly (phenylene ethers). Their processing is carried out at high temperatures (some-

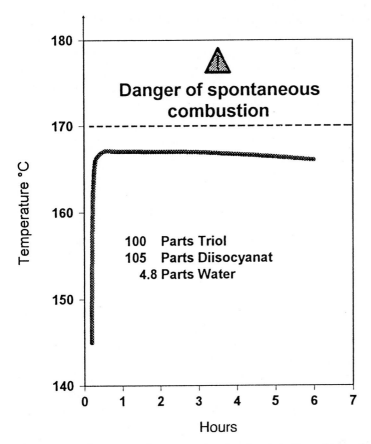

Fig. 1.7. Development of temperature in the interior of a polyurethane block (2×1×5 m, density: 21 kg/m^3)

times above 300 °C) and is thus at the border of possible thermolytic processes (pyrolysis). Degradation and crosslinking reactions can occur and discoloration is observed. the thermooxidative and thermomechanical degradation under processing conditions still needs extensive investigation.

1.4.9
Polymer Blends and Alloys

Mixing two or more polymers permits the preparation of so-called polymer blends with targeted properties, e.g. impact strength. Polyolefins with ethylene-propylene rubber, EPR, as "impact modifier" are gaining popularity and are used in a variety of applications such as automotive external parts, casings for utilities and similar construction parts.

Recently such blends have been produced by sequential polymerization in the reactor. the morphology is approximately the same as that obtained by compounding of the individual components in an extruder. Further well known impact resistant-modified plastics are based on styrene polymers, e.g. ABS, ASA, HIPS and MBS. Impact modified polymer blends using polyamides, polyester, polyvinyl chloride or polycarbonate with suitable rubber types find a variety of applications. All these polymer blends are multiphase systems, i.e. the individual components are incompatible with each other. With regard to thermal oxidation and thermo-mechanical degradation, blends behave during processing as the individual components alone. However, because of the complex rheological properties, significant shear forces may arise during processing.

1.5
Upper Limit of Processing Temperature

The highest possible processing temperature is determined by the possible chemical reactions caused by thermal degradation. This varies within broad limits from polymer to polymer. Polymers with defined ceiling temperature, T_C, decompose to a large extent back to the monomer. All other polymers break down into a variety of fragments of differing size. A detailed summary of the products resulting from thermal degradation of a multitude of polymers is listed in the Polymer Handbook [33].

1.6
Degradation of Polymers Under Long Term Thermal Conditions

In the course of their useful lifetime, polymers are subjected as end product to widely differing thermal stress. In applications such as, e.g., insulation of automotive hoods, temperatures up to $130\,°C$ are maintained over considerable periods of time. The polymer used should not undergo thermal damage during the lifetime of the vehicle. Substantial thermal stress can also also arise in applications outside automotive parts, in domestic appliances or in electronics. Table 1.6 lists the lifetime of some unstabilized polymers.

The thermooxidative stability of plastics depends very much on temperature. Because of the complex reactions of autoxidation, there is, so far, no definite relationship established between oven aging time

Table 1.6. Lifetime of some polymers after aging in a draft air oven

Polymer	Sample Thickness	Aging Temperature	Time to Failure	Failure Criteria
Polypropylene	1 mm	135 °C	<1 day	Embrittlement on bending
Polyethylene, HD	1 mm	120 °C	15 days	50% Retained tensile impact strength
Polyamide, 6.6	1 mm	120 °C	5 days	50% Retained tensile impact strength
Poly (butylene terephthalate)	1 mm	120 °C	7 days	50% Retained elongation

and service lifetime. Gugumus has shown, based on aging studies of polypropylene film, that evaluation of the experimental data according to *Arrhenius* does not produce a linear relationship [34] as shown in Fig. 1.8.

Gijsman et al. [35] have shown in the case of the thermal oxidation of polypropylene that two different reactive peroxides arise which decompose at different rates. The rise of the oxidation rate between 50 and 90 °C is explained by the rapid decomposition of thermooxidatively generated peracids. Such peracids can arise, e.g. by oxidation of intermediately formed aldehydes (Eq. 1.20), and might be involved in oxidation of numerous other polymers (Scheme 1.9).

Along with peracids, further oxidation products can be identified with the help of IR spectra (Fig. 1.9), showing that groups such as carbonyl, carboxylic acids, esters and γ-lactones [36] are formed by different oxidation reactions (Scheme 1.10).

Sample thickness and morphology substantially influence the behavior of a polymer with regard to its response to long term thermooxidative conditions. As already mentioned, oxygen is generally soluble only in the amorphous areas of partially crystalline polymers, e.g. in polyolefins. For this reason, oxidation of, e.g. polyolefins is heterogeneous with regard to the substrate and occurs predominantly where there are high peroxide concentrations [37–40]. Impurities or catalyst residues which are also found in these areas contribute further to locally high concentrations of oxidation products.

The polymer's available oxygen content is decisive for the way autoxidation proceeds. Oxygen is then consumed by different oxidation reactions. If by follow-up diffusion sufficient oxygen is steadily supplied (sample in equilibrium with oxygen saturation) then oxidation proceeds uniformly throughout the sample. If, however, oxidation reactions locally consume more oxygen than can be resupplied by diffu-

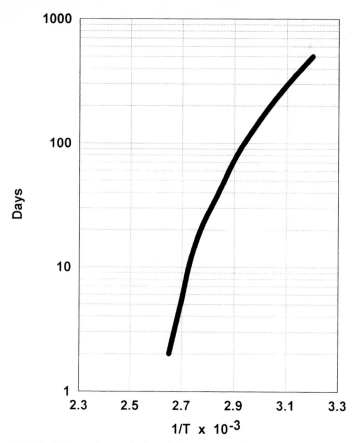

Fig 1.8. Aging of PP (film) at different temperatures, time to embrittlement

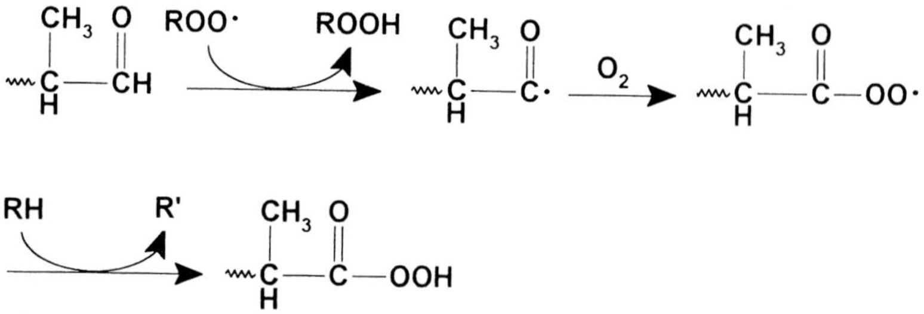

Scheme 1.9. Formation of peracids

sion into the sample's interior, then autoxidation slows down or can practically come to a stop.

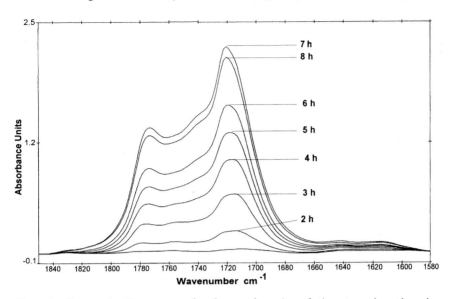

Fig. 1.9. Change in IR-spectra of polypropylene in relation to aging time in a draft air oven at 135 °C

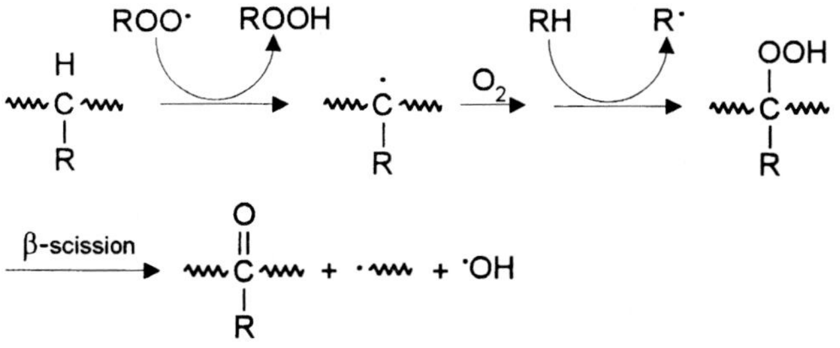

$$(1.21)$$

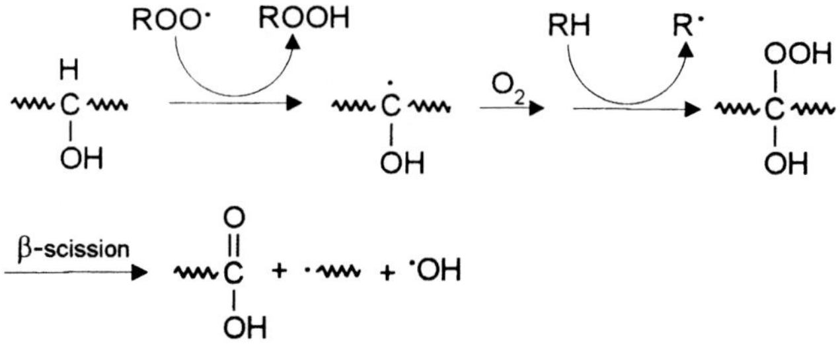

$$(1.22)$$

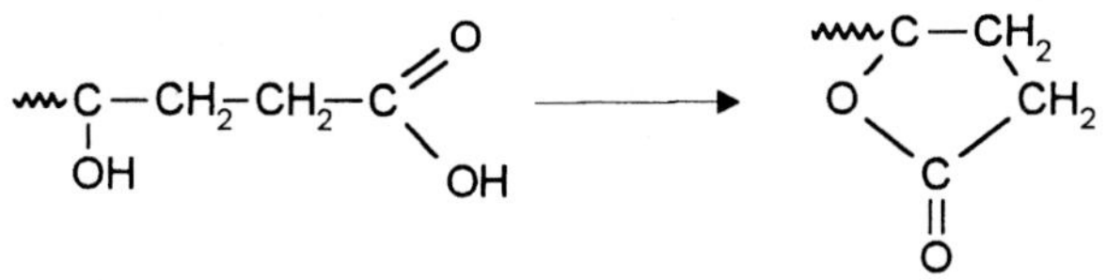

$$(1.23)$$

Scheme 1.10. Formation of oxidation products

Several authors [41–43] have developed kinetic models that can be used for the calculation of possible oxidation profiles (Fig. 1.10 and Fig. 1.11).

They are based on the assumption that *Fick's* second law applies to the oxygen diffusion into the polymer. The rate of change of oxygen concentration, C, at any point is

$$\frac{dC}{dt} = D\frac{d^2C}{dx^2} \tag{1.1}$$

where t is the time, x the depth of penetration perpendicular to the surface and D the diffusion coefficient of oxygen into the polymer. If the oxygen is consumed by autoxidation reactions at a rate r, Eq. (1.1) becomes modified to

$$\frac{dC}{dt} = D\frac{d^2C}{dx^2} - r. \tag{1.2}$$

Since the local reaction rate r is a function of the local oxygen concentration, $r \to r(C)$ and by assuming that the system reaches to a stationary state, $\frac{dC}{dt} = 0$, Eq. (1.2) can be written in the form

$$D\frac{d^2C}{dx^2} - r(C) = 0. \tag{1.3}$$

The oxidation profile can be calculated from Eq. (1.4), using appropriate boundary conditions. The concentration profile C(x) can thus be determined using this differential equation and appropriate boundary conditions.

The local conversion, $Q_{(x)}$ in the thick sample is

$$Q(x) = \int_0^t r(C)dt \tag{1.4}$$

and the average conversion, Q, for the whole sample

$$Q = \frac{1}{L}\int_0^L Q(x)dx. \tag{1.5}$$

In practice r(c) is rarely well defined – however it is possible to compute shapes of oxidation profiles for first-order (Fig. 1.10) and zero-order kinetics (Fig. 1.11) [44]. The computed depth profile is given by Q/Qs (relative conversion) and is plotted as a function of the normalized sample thickness, L. The thickness of the oxidised layer, TOL, is an increasing function of Φ^{-1} (see Eq. 1.6) which has the dimension of length.

For first-order kinetics, TOL is, as shown in Figure 1.10

$$TOL \cong \Phi^{-1} = \left(\frac{D}{k}\right)^{1/2} \qquad (1.6)$$

and for zero-order kinetics, Fig. 1.11

$$TOL = v\left(\frac{D}{k}\right)^{1/2} = v\Phi^{-1}. \qquad (1.7)$$

The variations of the thickness of the oxidised layer, TOL, with time and related to exposure conditions are of interest in view of the material's long term behavior. Measuring the average conversion Q over a sample's cross section using different experimental techniques (see Sects. 6.3.1, 6.4.2 and 6.5) gives insight into the boundary between no diffusion and diffusion controlled degradation.

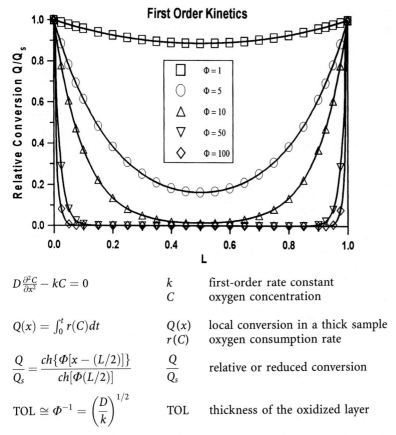

$D\frac{\partial^2 C}{\partial x^2} - kC = 0$	k	first-order rate constant
	C	oxygen concentration
$Q(x) = \int_0^t r(C)dt$	$Q(x)$	local conversion in a thick sample
	$r(C)$	oxygen consumption rate
$\dfrac{Q}{Q_s} = \dfrac{ch\{\Phi[x - (L/2)]\}}{ch[\Phi(L/2)]}$	$\dfrac{Q}{Q_s}$	relative or reduced conversion
$TOL \cong \Phi^{-1} = \left(\dfrac{D}{k}\right)^{1/2}$	TOL	thickness of the oxidized layer

Fig. 1.10. Computed shape of normalized oxidation profile for first order kinetics

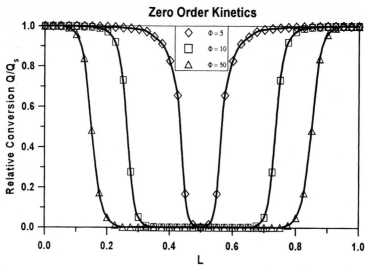

$$D\frac{\partial^2 C}{\partial x^2} - r_0 = 0; \ r_0 = kC_s \qquad \text{oxidation rate in the superficial layer}$$

$k = r_0/C_s$ Pseudo-first-order rate constant

C_s equilibrium oxygen concentration

$C = C_s + \frac{r_0}{2D}x(x - L)$

$C = C_s - \dfrac{r_0 L^2}{8D}$ concentration in the middle of the sample $(x = L/2)$

$C_c = C_s - \dfrac{r_0(2TOL)^2}{8D}$ C_c critical concentration
    TOL   thickness of the oxidized layer

$TOL = \left(\dfrac{2D(C_s - C_c)}{r_0}\right)^{1/2}$

$2(C_s - C_c) = v^2 C_s$ $0 \leq v \leq \sqrt{2}$
    $0 \leq C \leq C_s$

$TOL = v\left(\dfrac{Pp}{r_0}\right)^{1/2}$ $P = DS$ polymer permeability
    p oxygen pressure

$TOL = v\left(\dfrac{D}{k}\right)^{1/2} = v\Phi^{-1}$ $\Phi^{-1} = \left(\dfrac{D}{k}\right)^{1/2}$

$v = \left[2\left(1 - \dfrac{C_c}{C_s}\right)\right]^{1/2}$

Fig. 1.11. Computed shape of normalized oxidation profile for zero order kinetics

The critical film thickness for uniform oxidation depends on temperature and increases clearly with falling temperature. Oxygen diffusion is unlikely to play a role in film and fibre samples, but its effects are observed in thick sections.

If methods such as, e.g., γ-radiation or too high temperatures are used to accelerate aging of plastics specimens, then degradation caused by high reaction rate of oxidation and relatively low diffusion rate of the oxygen takes place predominantly in the surface layer of the polymer. This is the reason why too highly accelerated aging conditions can lead to erroneous conclusions with regard to aging behavior under normal conditions.

During aging of a plastic material under thermal stress, there is a correlation between increase of hydroperoxides concentration, [ROOH], and the loss of mechanical properties. Figure 1.12 shows schematically this relationship.

The concentration of hydroperoxides is too low at the beginning of thermooxidative degradation to be determined by titration methods. The time to a rise of measurable hydroperoxide concentration is referred to as induction period. With the steep rise of hydroperoxide concentration, there is a correlation between oxygen uptake of the sample and the oxidation products determined by IR spectroscopy (carbonyl absorption) and with polypropylene, the reduction of the average molecular weight of the polymer, M_w. Table 1.7 depicts the relationship between molecular weight decrease and aging of 1 mm thick unstabilized polypropylene plaques [20].

In partially crystalline polymers, with originally ductile characteristics, loss of mechanical properties is the result of molecular weight degradation caused by β-scission of the alkoxy radicals (Eq. 1.7, Scheme 1.1). The transition from ductile to brittle behavior is related to a critical molar mass, M_C. The loss of mechanical properties can be described with the help of fracture mechanics. Audouin et al. [44] summarized the relationship between some aspects of thermal aging, oxidation profile, and fracture mechanics.

In a sample displaying oxidative degradation on the surface ("skin core structure") micro cracks are formed in the course of diffusion-controlled aging. Crack propagation under stress can reach the ductile material or across the sample. Figure 1.13 is a photograph of such a crack starting from oxidized, brittle surface with an oxidation depth of approx. 150–200 µm, all the way far into the ductile material.

This crack formation was observed on a polyethylene pipe of medium density in a long term internal pressure test. Rappoport et al. [45]

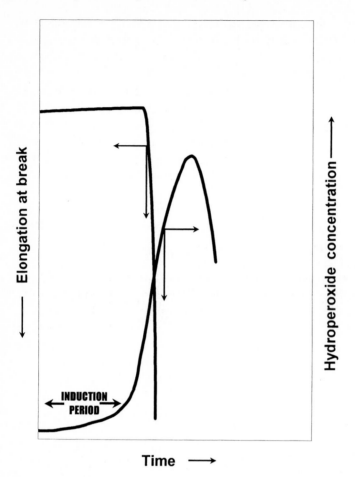

Fig. 1.12. Hydroperoxide concentration in relation to mechanical properties

investigated crack formation in polypropylene caused by thermal oxidation close to catalyst residues and under mechanical stress.

A summary of the above-mentioned findings shows:

(i)   the degradation of partially crystalline polymers, particularly polyolefins, under themooxidative aging conditions is a "heterogeneous process";

(ii)   the oxidation profile across a sample is determined by oxygen diffusion and the local consumption of oxygen, which is a function of the rate of oxidation reactions, and strongly influenced by thermal aging conditions;

Table 1.7. Change in molecular weight, M_w, molecular weight distribution, M_w/M_n, of unstabilized polypropylene[a] after oven aging at 135 °C

Hours	M_w	M_n	M_w/M_n
0	178,000	60,000	2.99
1	121,000	42,000	2.92
2	106,000	36,000	2.94
3	98,000	30,000	3.27

[a] Sample thickness 1 mm

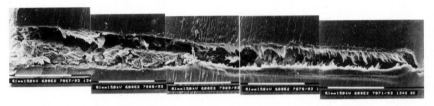

Fig. 1.13. Crack formation in PE-Md pipe. With permission, G. Dörner, Ph.D. Thesis, Montanuniversität Leoben, Austria

(iii) the "failure" of mechanical properties of the plastics material in the course of aging is also associated with crack formation on the surface and subsequent crack propagation under mechanical stress.

Besides the rate of diffusion of oxygen into the polymer matrix, the long term thermal stability of a polymer is governed by the ease of peroxide formation. Thus, the thermooxidative stability is linked to the rate of hydrogen abstraction from a C-atom in the main chain or a side group. Formation and decomposition of hydroperoxides can lead to complex oxidation processes as, e.g., the thermal degradation of polycarbonates [46].

Thermooxidative degradation of polycarbonate (Scheme 1.11) leads to strong yellowing of the polymer. Discoloration under long term thermal aging conditions is observed in other plastics, e.g. polyamides and polyester. In all these instances it should be taken into account that thermal aging is influenced by superimposed photochemical processes and possibly also by hydrolysis.

In polyacetals molecular weight degradation occurs, caused by the thermooxidatively initiated "deformylating reaction" generating formaldehyde.

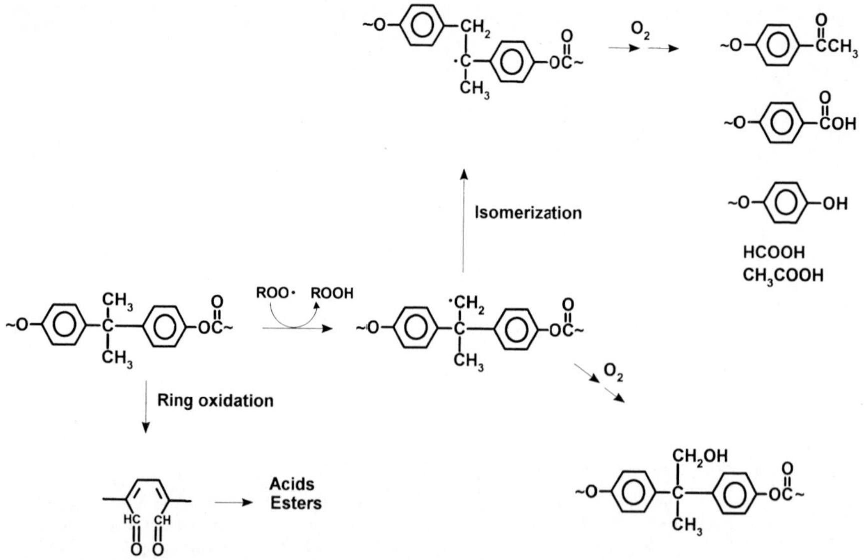

Scheme 1.11. Thermooxidative degradation of polycarbonate

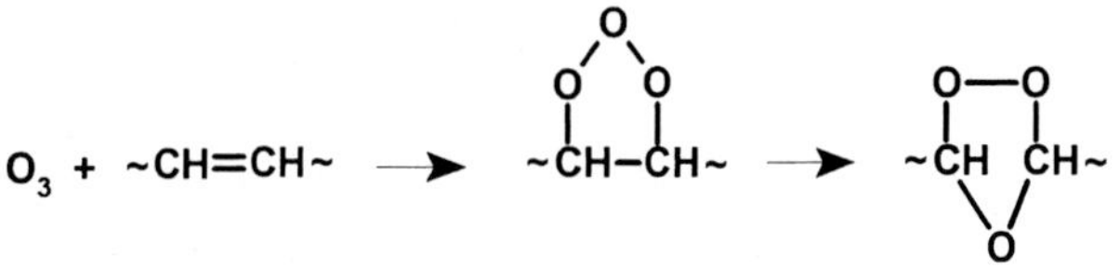

$$O_3 + \sim CH=CH\sim \longrightarrow \sim CH-CH\sim \longrightarrow \sim CH \quad CH\sim \qquad (1.24)$$

Scheme 1.12. Reaction of ozone with -C=C- double bonds

The degradation of elastomers with polydiene sequences is characterized by the formation of cracks on the surface of the rubber with subsequent crack growth under mechanical stress. In this instance, degradation may also involve the reaction of ozone with the -C=C- double bond and is connected, according to Eq. (1.24), with formation of ozonides (Scheme 1.12).

Elastomers based on largely saturated -C-C- polymer backbones such as EPR and EPDM degrade rather slowly under the effect of ozone. Kuczkowski [47] published a summary of current knowledge concerning ozonolysis and its inhibition.

1.7
Degradation of Polymers by Photooxidation

Degradation of polymers by UV light in the presence of air is another important physical factor in oxidation reactions. Figure 1.14 shows a

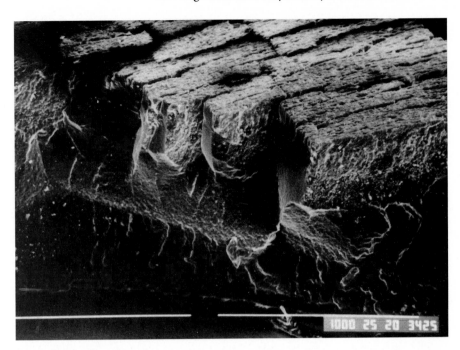

Fig. 1.14. SEM picture of the surface of a PP-EPR-copolymer (containing carbon black), after exposure to sunlight (Florida, 3 years)

SEM picture of the surface of a polypropylene-EPR-copolymer which was exposed to sunlight, showing the destruction of the polymer's surface.

With increasing temperature, thermooxidation may overlap photooxidation. Therefore factors influencing the oxidative deterioration of the polymers may contribute significantly to the overall degradation processes.

1.7.1
Absorption of Light by Polymers

The lower wavelength limit of sunlight reaching the earth's surface is about 300 nm.

Many of the commercially important polymers, e.g. polyethylene or polypropylene, should not absorb any sunlight since the longest wavelength absorption band for the polyolefins is in the region below 200 nm, caused by a σ–σ^* transition. The absorption of light by synthetic rubbers based on butadiene or isoprene copolymers is associated with a $\pi-\pi^*$ transition and occurs in the wavelength region of 180–240 nm. The longest absorption of polystyrenes is associated with the $\pi-\pi^*$ transition of the benzene ring and occurs in the wavelength region of 230–280 nm. The longest wavelength absorption band for polyacetals can be associated with a partially forbidden $n-\sigma^*$ transition below 200 nm. The longest wavelength absorption bands of polymers such as aliphatic polyamides, aliphatic polyesters or poly(meth)-acrylics are in the region below 200 nm, associated with an $n-\pi^*$ transition [48, 49].

The first fundamental law of photochemistry formulated by Grotthus and Draper states that only light absorbed by a molecule can be effective in producing photochemical change in the molecule. Polymers which do not absorb sunlight should not undergo photochemically induced reactions. In contrast to these polymers there are many commercial plastics which absorb sunlight owing to the chromophoric groups that form part of the polymer structure, e.g. aromatic polyesters and polyamides, polysulfides, poly (ether sulfones), polycarbonate and other polymers containing aromatic moieties.

Such polymers are prone to any of the photo-induced reactions known in the field of organic photochemistry.

The absorption spectra of some virgin polymers are shown in Fig. 1.15.

In any polymer which does not contain inherent chromophores, hydroperoxide groups, formed by thermal oxidation, will be present. Such hydroperoxide groups absorb in the UV region of sunlight and subsequent reactions lead to the formation of carbonyl groups. Catalysts used to produce the polymers, e.g. transition metals and Ziegler-Natta catalysts [50], are present in resins such as polyolefins. Atmospheric pollutants such as polycyclic aromatic hydrocarbons, absorbed by the polymer, may further contribute to the formation of components or chromophoric groups which may also absorb sunlight.

Formation of hydroperoxides by the reaction of molecular oxygen with alkyl radicals, $R^\bullet$, has been discussed in detail in the previous sections. Beside this reaction, singlet oxygen, $^1O_2$, can react in an "ene" -type reaction in the α-position to a -C=C- bond, leading to the formation of hydroperoxides (Eq 1.25, Scheme 1.13).

It is still a matter of discussion whether singlet oxygen plays an important role in oxidative degradation processes [51]. The triplet photosensitised production of singlet oxygen caused by carbonyl groups is a possible reaction and Gugumus [52] found improved photostability of polyethylene in the presence of 1,4-diazobicyclo[2, 2, 2] octane, DABCO, well known to act as a singlet oxygen quencher [53].

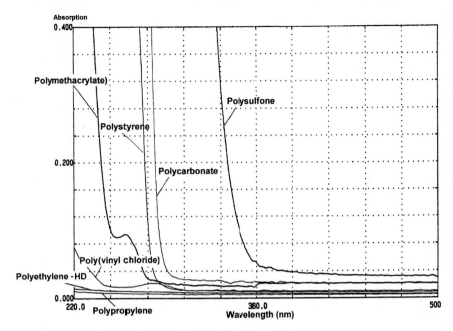

Fig. 1.15. Absorption spectra of commercial polymers

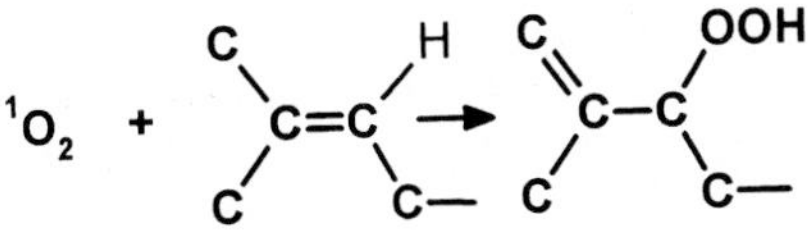

$$^1O_2 \; + \; \underset{C}{\overset{C}{\diagdown}}C=C\underset{C-}{\overset{H}{\diagup}} \longrightarrow \underset{C}{\overset{C}{\diagdown}}C-C\underset{C-}{\overset{OOH}{\diagup}} \qquad (1.25)$$

Scheme 1.13. Reaction of singlet oxygen with -C=C- double bonds

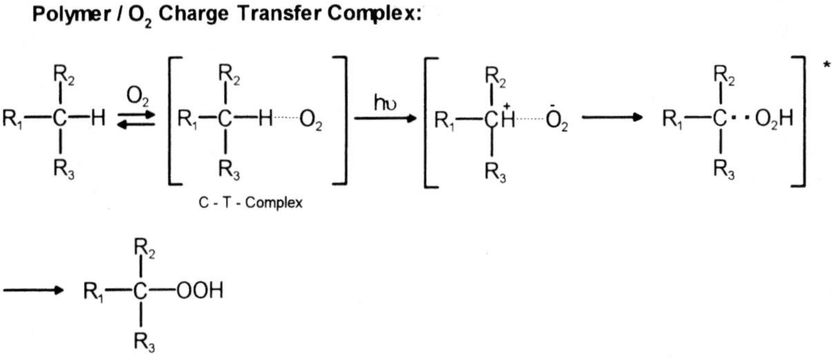

Polymer / O₂ Charge Transfer Complex:

$$R_1-\underset{R_3}{\overset{R_2}{\underset{|}{\overset{|}{C}}}}-H \;\rightleftarrows\; \left[R_1-\underset{R_3}{\overset{R_2}{\underset{|}{\overset{|}{C}}}}-H\cdots O_2 \right] \overset{h\upsilon}{\longrightarrow} \left[R_1-\underset{R_3}{\overset{R_2}{\underset{|}{\overset{|}{\overset{+}{C}}}}}H\cdots \dot{O}_2 \longrightarrow R_1-\underset{R_3}{\overset{R_2}{\underset{|}{\overset{|}{C}}}}\cdot\cdot O_2H \right]^*$$

C - T - Complex

$$\longrightarrow \; R_1-\underset{R_3}{\overset{R_2}{\underset{|}{\overset{|}{C}}}}-OOH \qquad (1.26)$$

Scheme 1.14. Formation of hydroperoxides via polymer/oxygen CT-complex

Formation of hydroperoxides by polymer/oxygen charge-transfer complexes or polymer/oxygen exciplex are mentioned in the literature [54, 55] (Scheme 1.14).

1.7.2
Photochemistry of Hydroperoxides

Hydroperoxides are the key intermediate in the breakdown of the polymer molecule. The photolytic decomposition of t-butylhydroperoxide in an inert solvent occurs at 313 nm with high quantum yield [56]. The photolysis of the hydroperoxide group yields a primary or secondary alkoxy radical, $RO^{\cdot}$. Subsequent β-scission would yield a ketone and an alkyl radical which will be transformed by further oxidation reactions into various oxidation products [57]. However, it is still a matter of investigations whether the β-scission reaction is important in photooxidation of polyolefins.

The photolysis of hydroperoxide groups under solar irradiation is a slow process, and the average lifetime of an -OOH group under constant irradiation is reported to be ~25 h [58–60]. Therefore, the most probable mechanism of photodecomposition of hydroperoxide groups is an energy transfer process from the excited carbonyl or aromatic

$$ROOH \xrightarrow{h\upsilon} \left[RO^{\cdot\cdot}OH\right]^* \longrightarrow RO^{\cdot} + {}^{\cdot}OH \qquad (1.27)$$

$$S_0 \xrightarrow{h\upsilon} \left[S\right]^* \xrightarrow{ROOH} S_0 + \left[RO^{\cdot\cdot}OH\right]^* \longrightarrow RO^{\cdot} + {}^{\cdot}OH \qquad (1.28)$$

Scheme 1.15. Sensitized photolysis of hydroperoxides

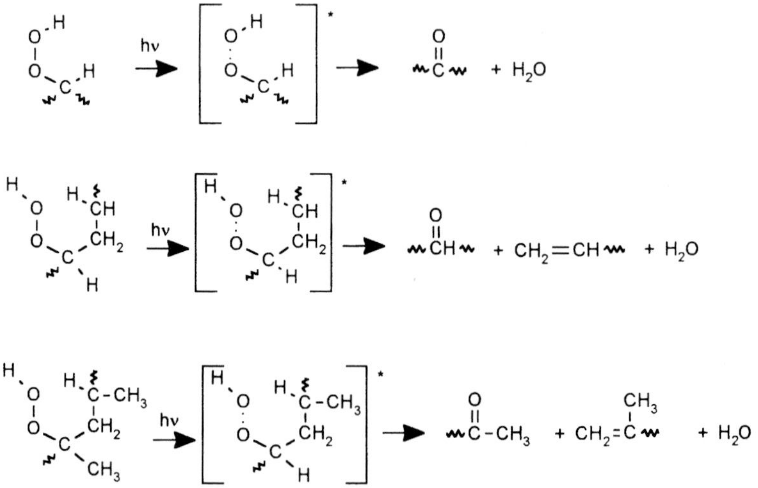

$$(1.29)$$

$$(1.30)$$

$$(1.31)$$

Scheme 1.16. Intramolecular decomposition of secondary and tertiary hydroperoxides

hydrocarbon group, acting as sensitizer, S, to the hydroperoxide groups as acceptors (Scheme 1.15).

Recently, intra- and intermolecular decomposition mechanisms based on the photolysis of the secondary and tertiary hydroperoxides as shown in Scheme 1.16 were proposed [61].

1.7.3
Photo-Induced Reactions

The classic photo-reaction in photooxidative degradation is the photolysis of hydroperoxides. However, photo-induced reactions are known in organic photo chemistry which lead to scission of the polymer backbone and thus to molecular weight degradation, e.g. Norrish Type I and Norrish Type II reactions (Scheme 1.17).

Photo-induced rearrangements such as the Photo-Fries reaction (Scheme 1.18) lead to changed structures of the polymer and completely change its properties.

Norrish Type I and Type II and photo-Fries reactions are referred to as photon-induced processes ("photolysis"). Photooxidative, as well as photolytic reactions are the basis of polymer degradation under the influence of light.

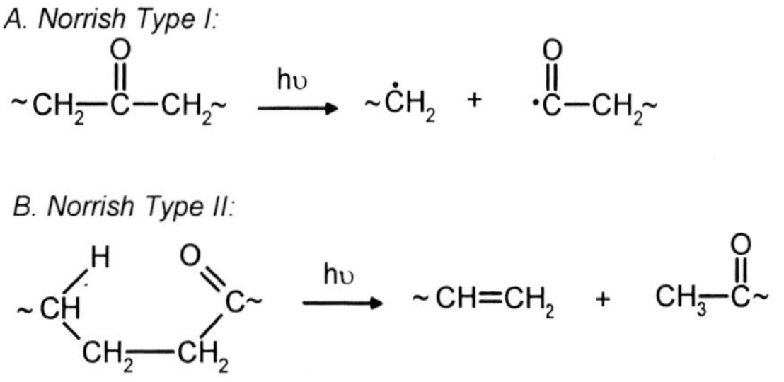

A. Norrish Type I:

$$\sim CH_2-\overset{\overset{\displaystyle O}{\|}}{C}-CH_2\sim \quad \xrightarrow{h\upsilon} \quad \sim\dot{C}H_2 \quad + \quad \overset{\overset{\displaystyle O}{\|}}{\cdot C}-CH_2\sim \tag{1.32}$$

B. Norrish Type II:

$$\sim \overset{\overset{\displaystyle H}{|}}{C}H \underset{CH_2-\overset{\displaystyle}{C}H_2}{\overset{\overset{\displaystyle O}{\diagdown}}{\diagup}} C\sim \quad \xrightarrow{h\upsilon} \quad \sim CH=CH_2 \quad + \quad CH_3-\overset{\overset{\displaystyle O}{\|}}{C}\sim \tag{1.33}$$

Scheme 1.17. Photo-induced Norrish Type I and Norrish Type II reactions

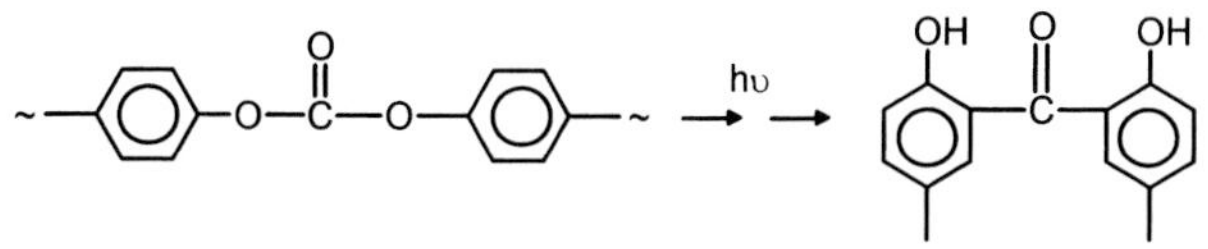

$$\sim\!\!-\!\!\bigcirc\!\!-O-\overset{\overset{\displaystyle O}{\|}}{C}-O-\!\!\bigcirc\!\!-\sim \quad \xrightarrow{h\upsilon} \quad \rightarrow \quad \underset{|}{\overset{OH}{\bigcirc}}-\overset{\overset{\displaystyle O}{\|}}{C}-\underset{|}{\overset{OH}{\bigcirc}} \tag{1.34}$$

Scheme 1.18. Photo-Fries rearrangement

1.7.4
Photo Degradation of Individual Substrates

1.7.4.1
Polyolefins

Upon irradiation of polypropylene plaques (thickness 1 mm) in a XENO 1200 exposure device for 160 h, the molecular weight decreases steadily with irradiation time [20], as shown in Table 1.8.

Table 1.8. Change in M_w and M_w/M_n of unstabilized polypropylene, after irradiation in a XENO 1200 exposure device, b.p. temp.: 55 °C

Hours	M_w	M_n	M_w / M_n
0	178,000	60,000	2.99
5	168,000	56,000	3.06
24	133,000	40,000	3.33
64	102,000	40,000	2.55

Sample form: 1 mm plaques

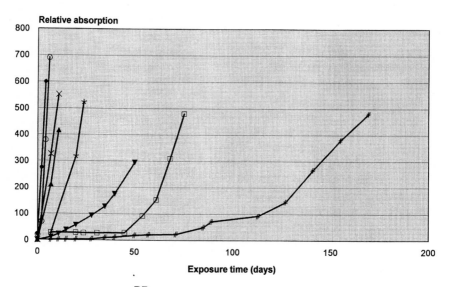

PP : ◆all UV ⊖320 nm ✳360 nm ⊕395 nm
PE-HD : ✳all UV ✦320 nm ▼360 nm ✦395 nm

Fig. 1.16. Build-up of carbonyl groups upon irradiation of a polyethylene and a polypropylene film, using different "cut-off" filters

The results clearly demonstrate that upon exposure of the polypropylene, chain scission reactions occur. Chain scission according to a Norrish Type I reaction as well as β-scission of alkoxy radicals leads to the formation of radicals which react with molecular oxygen, leading to further oxidation products according to the scheme of autoxidation. Analysis of the IR-spectra [20] shows the formation of further oxidation products, such as found in thermooxidative degradation, e.g. ketones, acids, esters, peracids, peresters and γ-lactones.

Figure 1.16 shows the formation of carbonyl groups upon irradiation of a polyethylene and a polypropylene film, using "cut-off" filters (320, 360 and 395 nm).

Polypropylene undergoes faster photooxidation compared with polyethylene because of the tertiary hydrogen atoms and the concerted mechanism of hydroperoxide formation.

1.7.4.2
Polystyrene

Polystyrene absorbs light only below 280 nm, and should, like polyolefins, not be degraded by sunlight. Analogous to polyolefins, the cause of polystyrene's photooxidative degradation is the photochemistry of hydroperoxides.

Chain scission leading to molecular weight degradation is the result of photolysis of the tertiary hydroperoxides [62, 63]. The resulting acetophenone structures act as sensitizers for the further photolysis of the hydroperoxides (Scheme 1.19). This explains polystyrene's sensitivity to photooxidation.

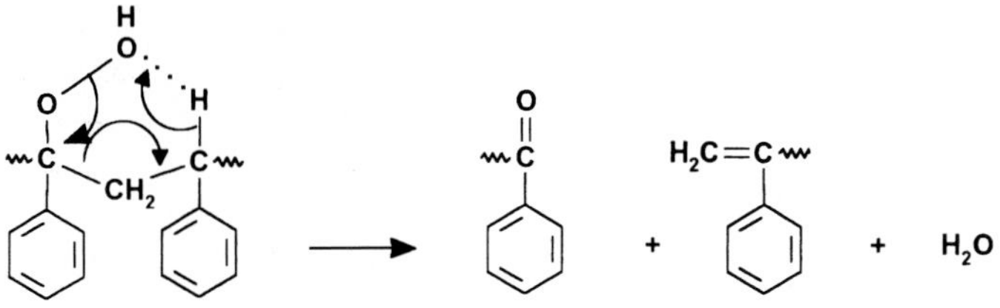

$$(1.35)$$

Scheme 1.19. Photooxidation of polystyrene

1.7.4.3
Polyester

The degradation of poly (ethylene terephthalate) and poly (butylene terephthalate) under the influence of light is already described in the

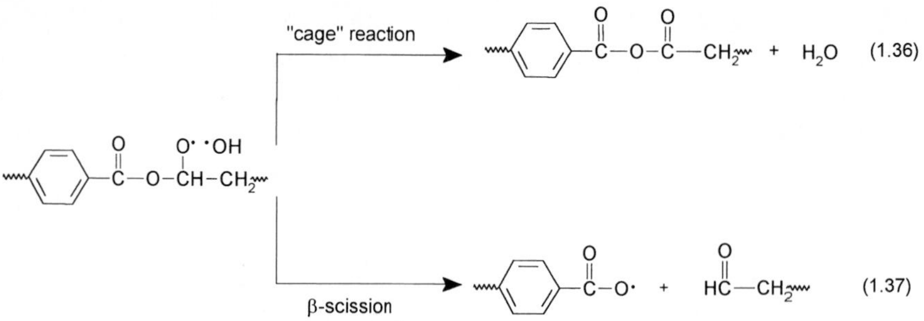

Scheme 1.20. Photooxidation of polyester

literature [64–68]. Photo induced reactions such as Norrish Type I and Type II, Eqs. (1.32) and (1.33), and photooxidative processes are observed (Scheme 1.20).

Furthermore, the formation of m-biphenyl structures is postulated [69] under photooxidative conditions.

1.7.4.4
Polyamides

Only a few investigations are known concerning the photooxidation of aliphatic polyamides. It can be assumed that photooxidative reactions such as the formation of hydroperoxide groups on the methylene group vicinal to the N-atom (amide bond) are the primary cause of photo-oxidative degradation. Photolysis of the hydroperoxide can lead subsequently to amide and carboxylic terminal groups [70] (Eq. 1.38, Scheme 1.21).

The formation of azomethine (Schiff's bases) which lead through further reactions to conjugated, unsaturated oligo(eneimine) structures, are said in the literature [71] to be the cause of discoloration.

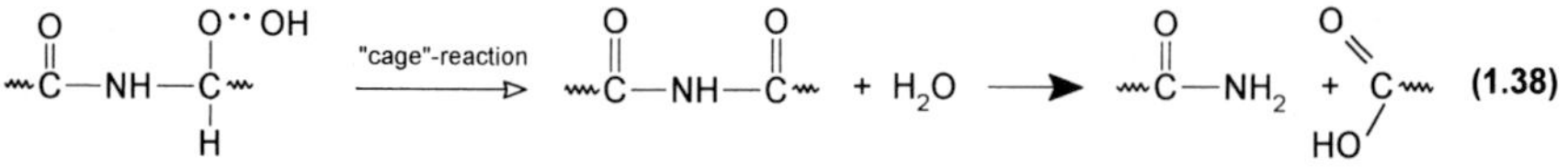

Scheme 1.21. Photooxidation of polyamides

1.7.4.5
Polyacetal

It is known that polyethers are relatively oxidation-sensitive. The abstraction of a hydrogen on the methylene group vicinal to the O-atom

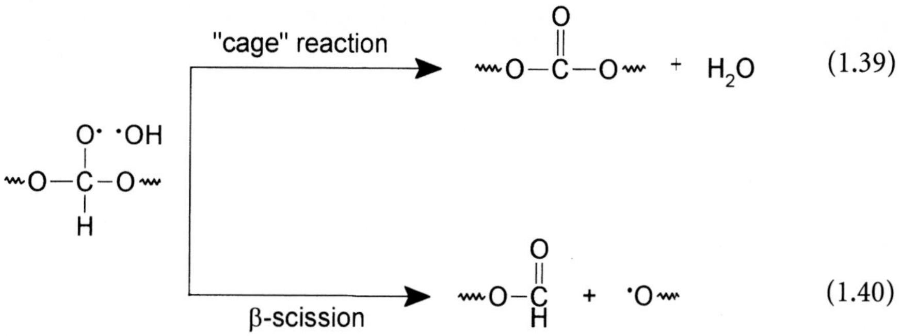

$$\text{O}\cdot \quad \cdot\text{OH}$$
$$\text{\large ---}O-\underset{\underset{H}{|}}{\overset{\overset{|}{}}{C}}-O\text{\large ---}$$

"cage" reaction $\longrightarrow$ $\text{---}O-\overset{\overset{\text{O}}{\|}}{C}-O\text{---}$ + H_2O $\qquad$ (1.39)

β-scission $\longrightarrow$ $\text{---}O-\overset{\overset{\text{O}}{\|}}{\underset{\underset{H}{|}}{C}}$ + $\cdot O\text{---}$ $\qquad$ (1.40)

Scheme 1.22. Photooxidation of polyacetals

(ether bond) subsequently yields a hydroperoxide group. Photolysis of the hydroperoxide in the "cage" leads to carbonate structures and water, and β-scission of the alkoxy radical leads to aldehyde groups [72], Eqs. 1.39 and 1.40 (Scheme 1.22).

Further oxidation results in the formation of acidic groups and, subsequently, to acid-catalysed depolymerisation of the polyacetal ("deformylation"). Weight loss and surface damage (chalking) occur with surface erosion [73–76].

1.7.4.6
Polycarbonate

Unstabilized bisphenol-A polycarbonate, when exposed to sunlight, undergoes discoloration, crosslinking occurs and cracks are formed on the surface. Since the PC polymer, by its structure, can absorb light in the region above 295 nm, photo-induced processes can occur.

BPA-PC undergoes chain scission reactions upon irradiation, leading to a decrease in molecular weight. Results of changes in molecular weight and discoloration are listed in Table 1.9.

Early work on the photoageing of BPA-PC suggested that the Photo-Fries reaction, as shown in Scheme 1.23, is the key mechanism [77, 78].

Recent investigations by Factor and co-workers [46] indicate that, upon exposure to sunlight, only small amounts of Photo-Fries products are formed. However, Lemaire et al. [80, 81] demonstrated that such products are easily photodegraded.

The photo-oxidation mechanism, promoted by light absorption in the region of 310–350 nm, caused by photo-Fries products, defects and impurities in the polymer [79], leads finally to coloured photo-products.

Table 1.9. Change in molecular weight, M_w, yellowness index, Y.I., and elongation of BPA-PC after irradiation in a 340 FL exposure device. Y.I., Yellowness Index, according to ASTM D 1925/70

Hours	M_W	Elongation %	Y.I.
0	29,900	96	3.00
285	25,000	42	4.54
535	24,550	26	6.54
800	22,000	15	8
970	18,800	0	11

Sample form: 25 µm film

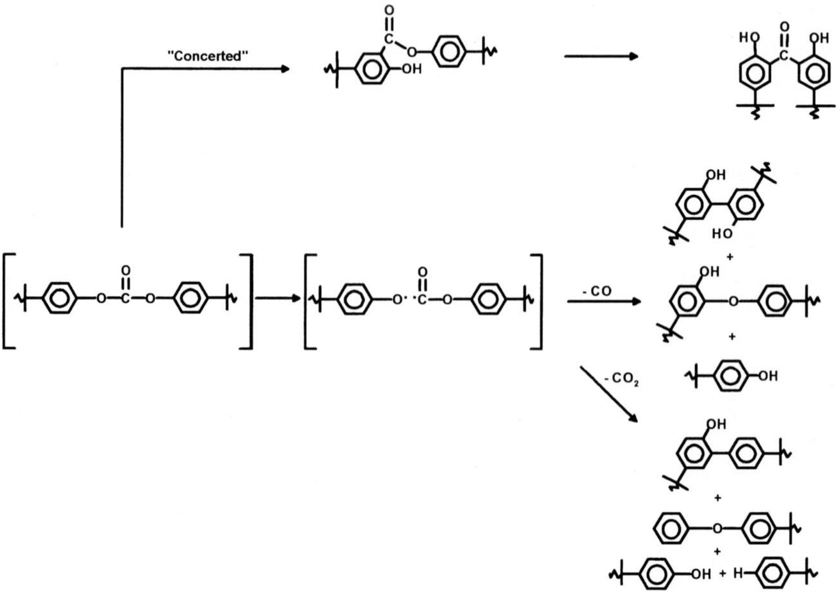

Scheme 1.23. Photo-Fries and photodegradation reactions involving bisphenol-A polycarbonate

The main degradation products can be assigned to side chain oxidation products, initiated by thermooxidative and photooxidative degradation processes.

1.7.4.7
Polyurethanes

The most technically important polyurethanes, the aromatic PUR, contain either polyester or polyether segments. the mechanism of photo degradation leading to loss of mechanical properties and discoloration of the substrate are photolytic reactions such as the photo-Fries rearrangement and photooxidative processes [82–85]. This is analogous to photo degradation of polyester and polyether.

1.7.4.8
High Performance Engineering Thermoplastics

High performance engineering thermoplastics are thermally stable polymers with aromatic structures in the polymer chain such as poly-sulfones, poly (ether sulfones), poly (phenylene sulphides), poly (ether imides), poly (aryl-ether ketones) and poly (phenylene ether). Polymers with such structures, therefore, absorb UV light in the long wave section of the spectrum. However, photolysis and photooxidative degradation of these polymers needs further elucidation. Some investigations have been published regarding photo degradation of poly (phenylene ether) [86, 87], polysulfones [88, 89], aramids [90], and poly (amide imides) [91].

1.7.4.9
Polymer Blends and Alloys

Photo degradation of polymer blends corresponds largely to that of the individual components. Detailed studies are lacking and synergistic as well as antagonistic effects cannot be excluded if the phases absorb light in differing ways.

1.7.5
Oxidation Profile of a Sample's Cross-Section

In many polymers, photooxidation occurs preferentially on the surface. Carlsson and Wiles [92] have shown that photooxidation of polypropylene takes place on the surface to a depth of 5 μm with subsequent micro crack formation. the behaviour of the aged product is determined by the fracture mechanics of crack propagation [93]. Crack formation on the surface manifests itself first as "chalking", i.e. light is reflected from the surface in different ways. Figure 1.17 shows the sur-

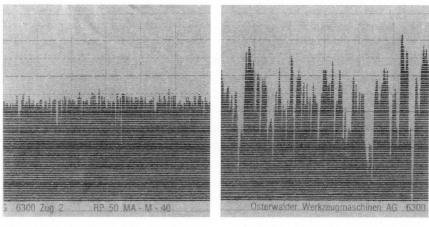

Fig. 1.17. Surface roughness, RA, (magnifying ratio 10 000 x) before and after aging of PP homopolymer in a Weather.O.Meter exposure device

face (sample cross section) of a PP-homopolymer plaque before and after light exposure in a Weather.O.Meter, W.O.M., exposure device.

For this reason, protection of the polymer's surface, even of filled or pigmented materials by means of suitable light stabilizers, is particularly important.

CHAPTER 2

Principles of Stabilization

2.1
Stabilization Against Thermo-Oxidative Degradation

2.1.1
Inhibition of Autoxidation

Stabilizers are chemical substances which are added to polymers in small amounts, in general, at most 1–2 wt% and are capable in the course of autoxidation of trapping emerging free radicals or unstable intermediate products such as hydroperoxides and to transform them into stable end products.

Within the cycle of autoxidation, the following are possible paths are shown as Fig 2.1.

Metallic impurities such as titanium, chromium, aluminium or iron, mainly as the corresponding ions, M^{n+}, are a source for the formation of alkyl radicals under processing conditions and during the life cycle of the end product. These metals or ions originate mainly from the

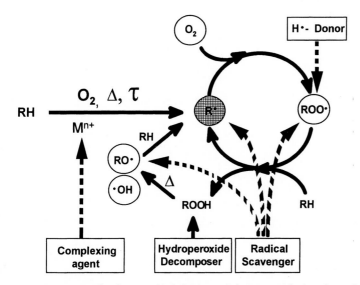

Fig. 2.1. General scheme of inhibition of thermooxidative degradation

catalysts used for the manufacture of polymers, particularly polyolefins. Suitable deactivation of the "active" form of these catalysts is, therefore, mandatory after polymerization. Generally, this deactivation is carried out by treating the polymer with moist nitrogen, steaming, or by the addition of alcohols immediately after polymerization. Because of the very low concentrations of such catalyst residues, the addition of suitable complexing agents as stabilizers is usually not effective.

In applications of polymers in direct contact with metals, such as cables with copper conductors and a polymer as insulator, the use of metal deactivators, MD, imparts significant improvement of the lifetime of such materials.

In principle, scavenging of the primary macroalkyl radicals, $R^•$, would immediately stop autoxidation. However, the rate of reaction of molecular oxygen k_2 (Eq. 2.1) is so high, $\sim 10^7 - 10^9$ mol^{-1} s^{-1} that this reaction can hardly be avoided (Scheme 2.1).

Under oxygen-deficient conditions, the use of alkyl radical scavengers can contribute significantly to the stabilization of the polymer. Such stabilizers are referred to as chain breaking acceptors [CB-A].

The rate-determining step in the course of autoxidation is the abstraction of a hydrogen from the polymer backbone (Scheme 2.2, Eq. 2.2), by the peroxy radical, $ROO^•$, thus forming the hydroperoxide, ROOH. If the peroxy radical is offered a substantially more easily abstractable hydrogen by a suitable hydrogen donor, InH, then the reaction meets competition according to Eq. (2.3), i.e. k_a [InH] > k_3 [RH].

$$R^• + O_2 \xrightarrow{\ k_2\ } ROO^• \qquad (2.1)$$

Scheme 2.1. Reaction of molecular oxygen with an alkyl radical

$$ROO^• + RH \xrightarrow{\ k_3\ } ROOH + R^• \qquad (2.2)$$

$$ROO^• + InH \xrightarrow{\ k_a\ } ROOH + In^• \qquad (2.3)$$

$$In^• + RH \xrightarrow{\ k_{ct}\ } InH + R^• \qquad (2.4)$$

Scheme 2.2. Inhibition reactions involving H-donors

H-donors (designated as InH here) are known as chain breaking donors [CB-D]. Suitable H-donors are characterized by the fact that they do not react further by abstraction of a hydrogen from the polymer backbone, according to Eq. (2.4), i.e. k_{ct} is small.

Scavenging of the $RO^{\bullet}$ and $HO^{\bullet}$ radicals, which are far more reactive than the peroxy radicals, $ROO^{\bullet}$, is not possible practically by using radical scavengers. For this reason, to avoid chain branching during autoxidation according to Eq. (1.9), so-called hydroperoxide decomposers, HD, are used as co-stabilisers. They decompose hydroperoxides forming "inert" reaction products.

Stabilizers of the type CB-D and CB-A are also referred to as primary antioxidants. Hydroperoxide decomposers, HD, are classified as secondary antioxidants. The "art" of stabilizing polymers consists of the appropriate choice of stabilizers and stabilizer blends for a given substrate.

In the following we discuss the way in which the various stabilizer types function and subsequently their use for the stabilization of different polymers during processing and the lifetime of the end product.

2.1.2
H-Donors and Radical Scavengers

2.1.2.1
Phenolic Antioxidants

The fact that phenols and amines are capable of highly efficiently inhibiting autoxidation was recognized long ago. Scott summarized these investigations in his book "Atmospheric Oxidation and Antioxidants" [94, 95]. Furthermore, the chemistry of inhibiting autoxidation by means of phenols has been investigated exhaustively by Pospisil [96, 97] and Henman [98].

Phenolic antioxidants are the most widely used stabilizers for polymers. Pospisil et al. [99–101] have summarized the state of the art regarding the mechanisms of antioxidant action. The key reaction in the stabilization of polyolefins by phenolic antioxidants is the formation of hydroperoxides by transfer of a hydrogen from the phenolic moiety to the peroxy-radical resulting in the phenoxyl-radical according to Eq. (2.5), Scheme 2.3.

The steric hindrance by substituents, e.g. *tert*-butyl groups in the 2- and/or 6-position, influences the stability of the phenoxyl-radical or the mesomeric cyclohexadienonyl-radicals. Sterically hindered phenols can be classified according to the substituents' 2-, 4-, and 6-position.

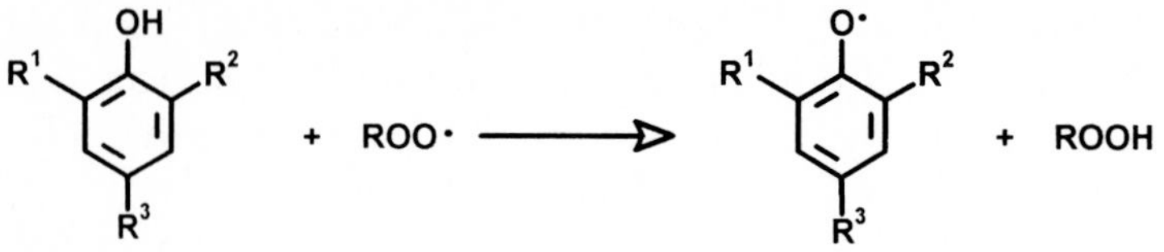

Scheme 2.3. Transformation of peroxy radicals into hydroperoxides by hindered phenols

Table 2.1. Antioxidant efficiency of some 2,4,6-trialkylphenols at 60 °C

H - DONOR	R^1	R^2	R^3	Antioxidant efficiency [a]	Chain transfer [b]
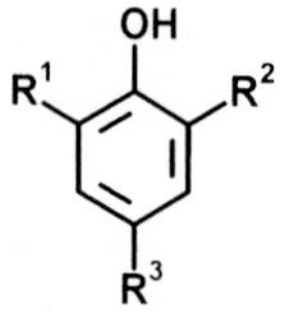	H	H	t-Butyl	80 - 100	0.0086
	Methyl	Methyl	t-Butyl	16 - 31	Large
	t-Butyl	t-Butyl	t-Butyl	32	0

a) k_a / k_3

b) k_{ct}

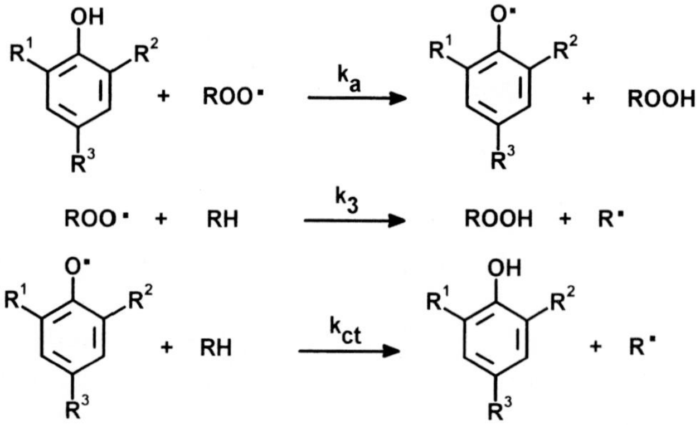

The efficiency of an antioxidant as a function of the substituents was investigated by Bickel and Kooyman. Some of the results are listed in Table 2.1 [102].

The results show that sterically hindered phenols are capable of preventing the abstraction of a hydrogen from the polymer backbone according to Eq. (2.6). The reactivity of the formed phenoxyl radical is significantly influenced by the substituents in 2- and 6-position. Bulky substituents prevent the reaction of the phenoxyl radical with the polymer (Eq. 2.7) and suppress dimerization of two phenoxyl radicals.

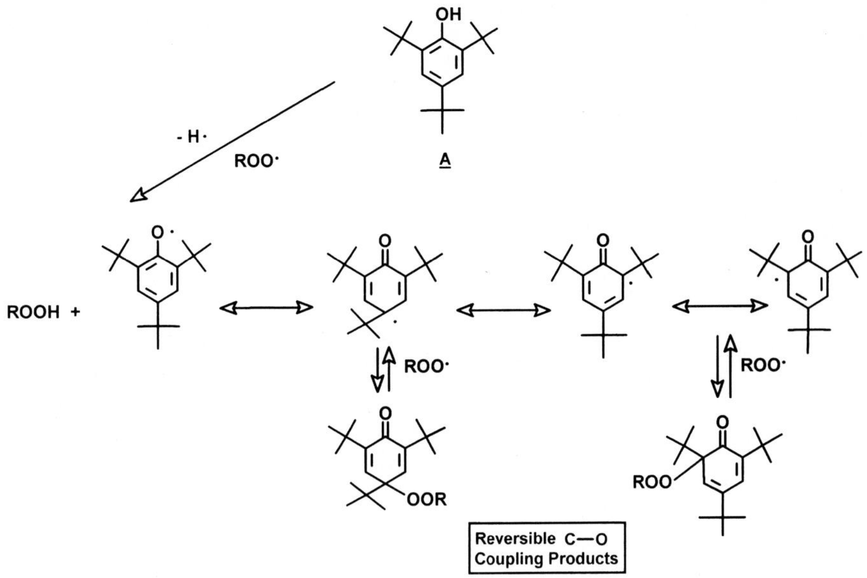

Scheme 2.4. Reactions involving fully sterically hindered phenols *A*

The rate of hydrogen abstraction from phenol increases with decreasing steric hindrance in 2- and 6-position.

Sterically hindered phenols are not only effective as H-donors. They are able to undergo numerous further chemical reactions that contribute to the inhibition of autoxidation. Sterically hindered phenols can be classified as follows.

"Fully sterically hindered phenols" *A*: (Scheme 2.4) substituents in 2-, 4- and 6-position having no H-atom on the C-atom vicinal to the phenyl group (no tautomeric benzyl-radical formation possible). The contribution of such phenols to stabilization consists essentially of the stoichiometric reaction between the phenol and the peroxyl-radical. The cyclodienonyl-radicals can add to ROO•-radicals (Scheme 2.4), although this reaction is reversible.

"Partially hindered phenols" such as *B*, and *C*: (Schemes 2.5 and 2.6) substituents at least in 4- (or 2- or 6-position) having H-atoms on a C atom vicinal to the phenyl group. The original phenol *B* or *C* is reformed by a disproportionation reaction, resulting in the corresponding quinonemethides. Inter- and intramolecular recombinations lead to generally irreversible C-C coupling products.

Reactions between oxygen-centred and carbon-centred radicals lead to reversible C-O coupling products.

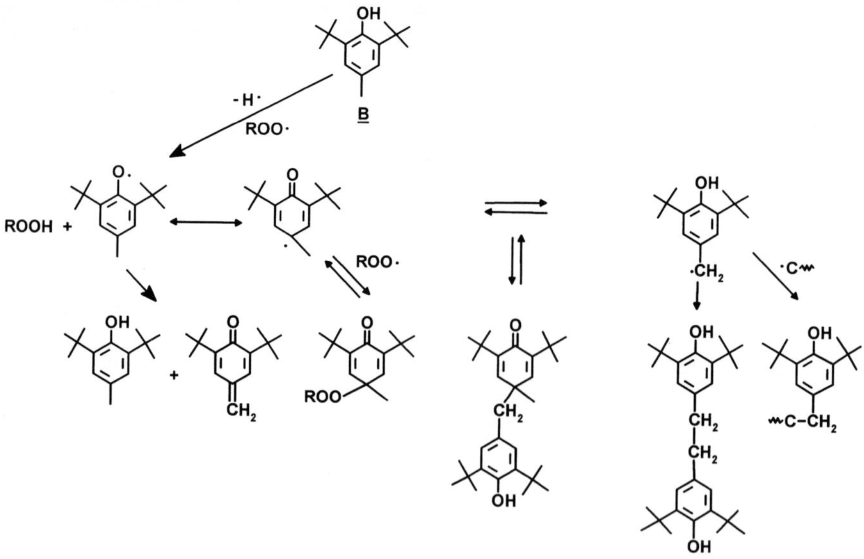

Scheme 2.5. Reactions involving partially sterically hindered phenols *B*

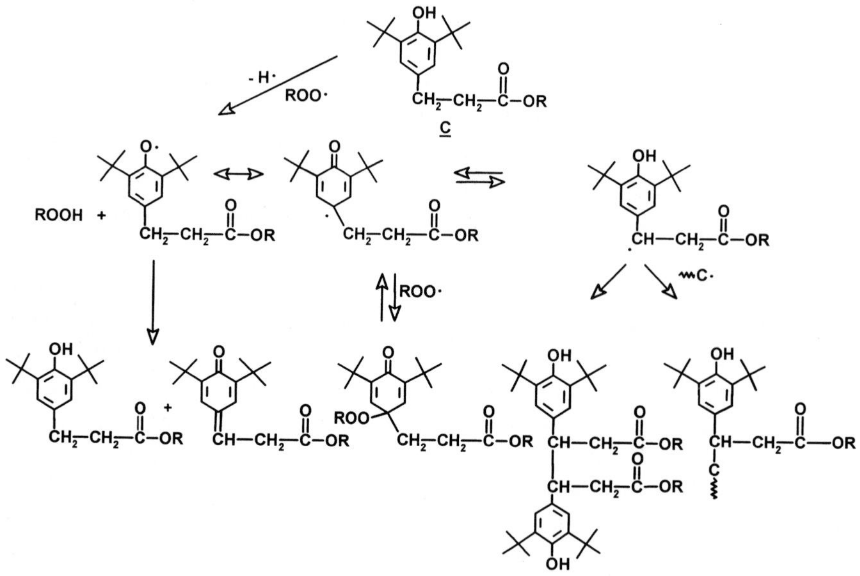

Scheme 2.6. Reactions involving partially sterically hindered phenols *C*

Through the disproportionation reaction of the respective phenoxyls of type *B* and *C* quinonemethides can be formed. The quinonemethides react with alkyl, alkoxy and peroxy radicals [103]; (Scheme 2.7).

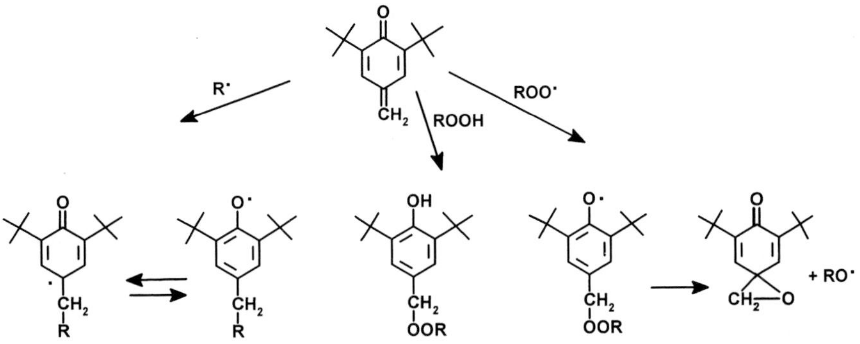

Scheme 2.7. Reactions involving quinone methides

Substituents in 2-, 4-, and 6-position are electronically equivalent. For this reason all corresponding reactions are also valid for these positions.

Sterically hindered phenols used currently on a commercial scale are generally of the substituent class of type C, i.e. they have a propionate group in the 4-position.

Such phenols contribute to stabilization by stepwise reactions resulting in stable transformation products in an "over-stoichiometric" way (related to the equivalents of available phenolic groups). In this case, reference is made to a stoichiometric factor f, larger than one [104, 105]. In any event, phenolic antioxidants are thus consumed.

2.1.2.2
Aromatic Amines

Secondary aromatic amines, and particularly aromatic diamines, are extremely efficient H-donors. Their function is based on the following reaction (Scheme 2.8, Eq. 2.8).

Bickel and Kooyman [106] investigated the relative antioxidant efficiency of some aromatic amines by measuring initial rates of oxidation using a standard hydrocarbon. The results are listed in Table 2.2.

$$\langle \_ \rangle\text{-NH-R' + ROO}^\bullet \longrightarrow \langle \_ \rangle\text{-}\dot{\text{N}}\text{-R' + ROOH} \qquad (2.8)$$

Scheme 2.8. Transformation of peroxy radicals into hydroperoxides by aromatic amines

Table 2.2. Antioxidant efficiency of some aromatic amines at 60 °C

H - Donor	Antioxidant efficiency [a]	Chain transfer [b]
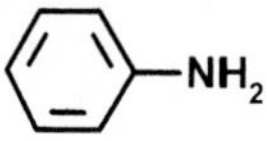 ⬡—NH_2	~ 40	~ 0.07
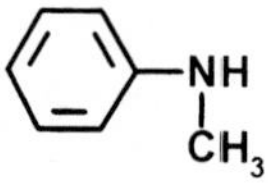 ⬡—NH—CH_3	15	0.03
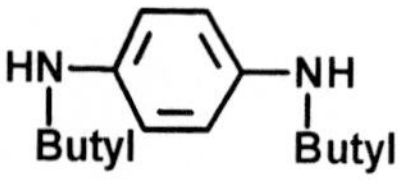 HN(Butyl)—⬡—NH(Butyl)	> 10,000	--

a) k_a/k_3
b) k_{ct}

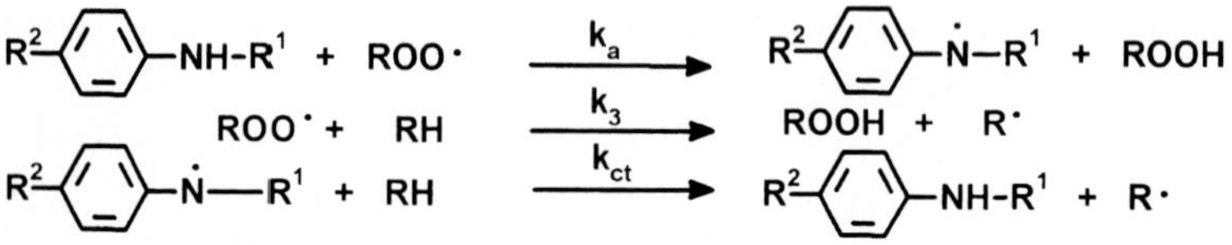

$$R^2{-}⬡{-}NH{-}R^1 + ROO^\cdot \xrightarrow{k_a} R^2{-}⬡{-}\overset{\cdot}{N}{-}R^1 + ROOH \quad (2.9)$$

$$ROO^\cdot + RH \xrightarrow{k_3} ROOH + R^\cdot \quad (2.6)$$

$$R^2{-}⬡{-}\overset{\cdot}{N}{-}R^1 + RH \xrightarrow{k_{ct}} R^2{-}⬡{-}NH{-}R^1 + R^\cdot \quad (2.10)$$

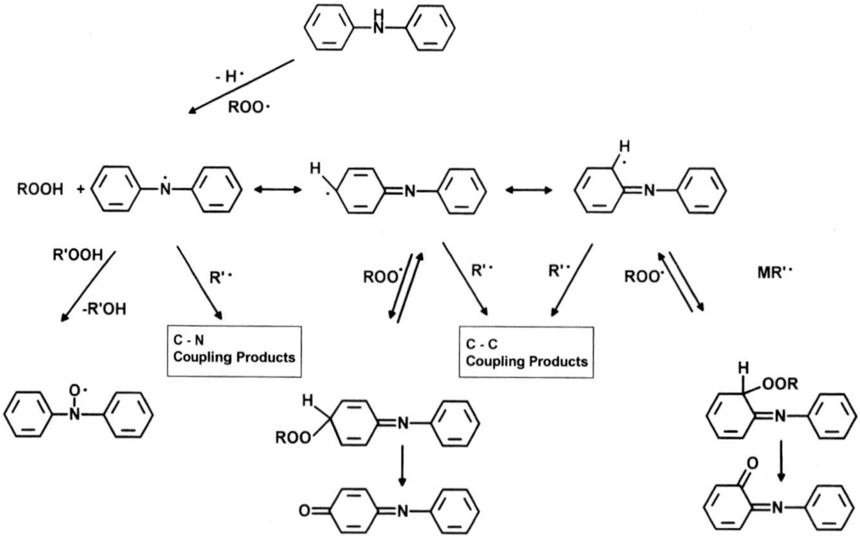

Scheme 2.9. Reactions involving aromatic amines

The primary reaction products can subsequently react like the phenols in further transformation steps forming various coupling products, as shown in Scheme 2.9 for diphenylamine.

The use of amine stabilizers can lead to pronounced discoloration of the substrate caused by the transformation products of the aromatic amines. For that reason, their application is essentially limited to carbon black filled rubbers (vulcanizates) and elastomers and other substrates where discoloration is not a critical issue.

2.1.2.3
Sterically Hindered Amines

Sterically hindered amines, designated "Hindered Amine Stabilizer, HAS", are effective stabilizers against thermal degradation of polyolefins [107, 108]. The activity of these amines as antioxidants is based on their ability to form nitroxyl radicals. The reaction rate of nitroxyl radicals with alkyl radicals appears to be only slightly lower than that of alkyl radicals with oxygen (Table 2.3), [109, 110].

For this reason, nitroxyl radicals are efficient alkyl radical scavengers. The mechanism of nitroxyl radical formation from sterically hindered amines is a controversial issue. Diverging opinions are recorded in the literature concerning the function of nitroxyl radicals. The reaction of an alkyl radical with the N-O$^\bullet$ radical leads to the formation of the hydroxylamine ether, NOR [111, 112]. This, in turn, reacts with a peroxy radical, ROO$^\bullet$, resulting in ROOR and the reformation of the nitroxyl radical as shown in Scheme 2.10.

Scott et al. [113] proposed a suitable reaction path (Scheme 2.11).

Table 2.3. Rate constants for the reactions of O_2 and nitroxylradical with alkyl radicals at 20 °C (in mol^{-1} s^{-1})

R$^\bullet$	O_2	
	2.7×10^9	$\sim 0.5 \times 10^9$
$(CH_3)_3C^\bullet$	5×10^9	$\sim 1 \times 10^9$

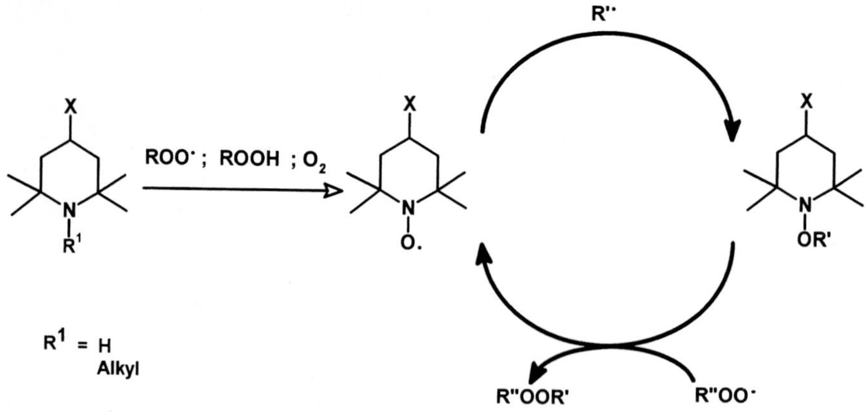

Scheme 2.10. Stabilization mechanism of hindered amines ("HAS"), I

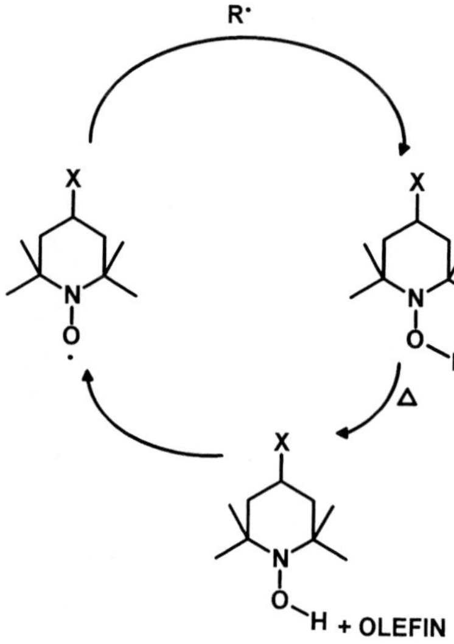

Scheme 2.11. Stabilisation mechanism of hindered amines ("HAS"), II

Felder et al. [114] have shown that peroxy radicals can oxidize sterically hindered amines to nitroxyl radicals. Turro et al. [115–117] found that the yields of nitroxide in the reaction of HAS with alkylperoxy radicals are low relative to acylperoxy radicals. The mechanism for this reaction is shown in Scheme 2.12.

The results are congruent with Gijsman's [118] findings concerning the effects of sterically hindered amines as long term thermal stabilizers. The reaction of the acylperoxy radical R(O)OO• with the steri-

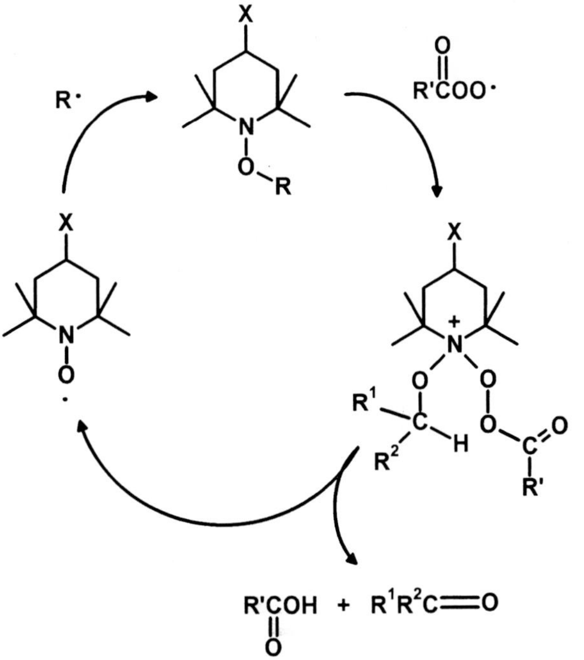

Scheme 2.12. Stabilisation mechanism of hindered amines ("HAS"), III

cally hindered amine or the nitroxy radical forming "inert" reaction products contributes significantly to thermooxidative stabilization. The postulated processes proceed cyclically while reforming the nitroxy radical until competing reactions destroy the sterically hindered amine by opening the ring.

It should be borne in mind, however, that the nitroxyl radicals are formed in the course of the polymer's autoxidation. For the stabilization of the polymer it is, therefore, necessary that processing of the melt requires an appropriate melt stabilization.

2.1.2.4
Hydroxylamines

Recently it was found that hydroxylamines can contribute in various ways to the stabilization of polymers [119]. The postulated reactions are summarized in Scheme 2.13.

The reactive species is the intermediary nitrone which is capable of scavenging C-radicals under oxygen-deficient conditions. Scott and Chakraborty [120] have shown that nitrones are active antioxidants.

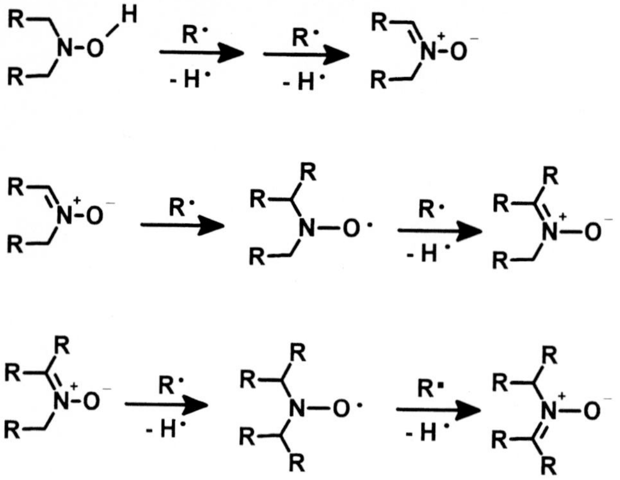

$$(2.11)$$

$$(2.12)$$

$$(2.13)$$

Scheme 2.13. Reactions involving nitrones and hydroxyl amines

2.1.2.5
C-Radical Scavengers

As already mentioned, trapping the primarily formed C-radicals would interrupt the chain reaction of autoxidation. Because of the high reaction rate of C-radicals with molecular oxygen, this concept can be considered mainly under low oxygen concentrations, i.e. during processing of a polymer as melt in an extruder. So far, few alkyl radical scavengers are mentioned in the literature as effective stabilizers. Scheme 2.14 depicts an effective principle developed by Yachigo et al. [121–123]:

The macro alkyl radical $R^{\bullet}$ adds onto the -C=C- bond of the acrylate group. Intramolecular shift of the hydrogen from the phenol group to the carbonyl of the ester group subsequently results in the stable phenoxyl radical. Such stabilizers are very effective in styrene-butadiene-styrene copolymers, preventing crosslinking, or in styrene-isoprene-styrene copolymers against degradation during processing (see Sect. 3.4.3).

Other compounds effectively scavenging C-radicals under low oxygen concentrations mentioned in the literature are benzofuranone derivatives [124, 125], as shown in Scheme 2.15.

Because the autoxidation proceeds sequentially, starting with the alkyl radical, $R^{\bullet}$, to the peroxy radical, $ROO^{\bullet}$, and subsequent formation of the hydroperoxide, $ROOH$, it is evident that one single functional

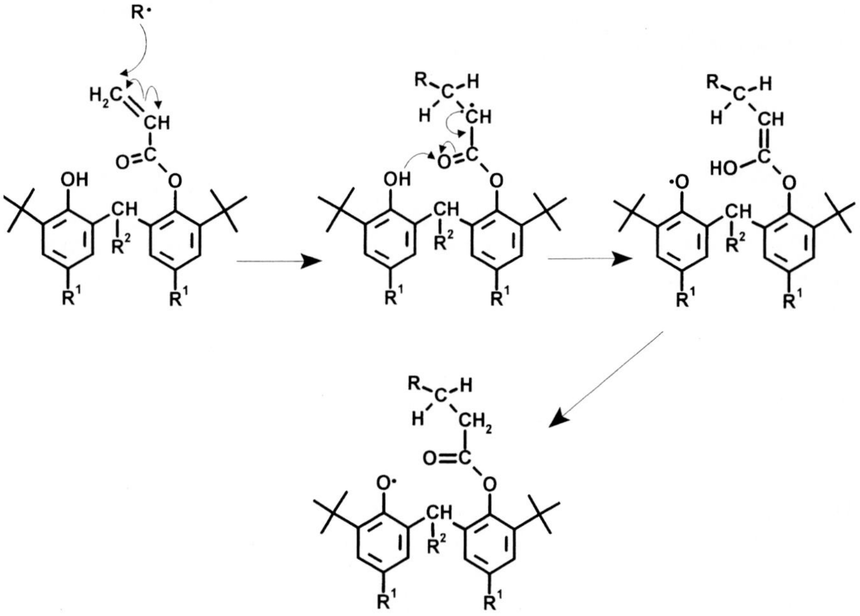

Scheme 2.14. Alkyl radical scavenging involving intramolecular hydrogen shift

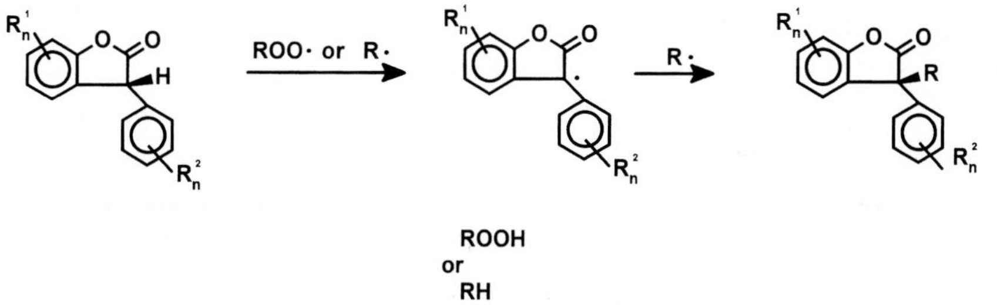

Scheme 2.15. Alkyl radical scavenging involving benzofuranone derivatives

principle is not sufficient for the inhibition of the chain reaction. For this reason, stabilizers are used in practice that react in different ways with the species formed in the course of autoxidation. There are stabilizers with groups in the same molecule that contribute in different ways to the inhibition of autoxidation. On the other hand, the same effect can be achieved by the use of suitable stabilizer combinations.

2.1.3
Hydroperoxide Decomposers

Hydroperoxides arising in the autoxidation chain reaction are the source for further oxidative degradation of polymers.

The extremely reactive alkoxy and hydroxy radicals, RO$^\bullet$ and $^\bullet$OH, result from the homolytic split of hydroperoxides. The rate of this reaction depends strongly on temperature and the hydroperoxide in question, although at processing temperatures of thermoplastic polymers (above 200 °C) it is relatively high. It follows that a hydroperoxide decomposer has to be able to compete with this reaction.

2.1.3.1
Phosphites and Phosphonites

Organophosphorus compounds of trivalent phosphorus were found to be efficient hydroperoxide decomposers. Schwetlick [126] thoroughly investigated the reactions of various phosphites and phosphonites. Some results regarding the efficiency in decomposing cumyl hydroperoxide at 30 °C are listed in Table 2.4.

A variety of phosphites and phosphonites are mentioned in the literature [127]. A hydroperoxide reacts exactly stoichiometrically forming the corresponding alcohol, simultaneously oxidising the phosphite to the corresponding phosphate (Scheme 2.16).

Table 2.4. Rate constants of phosphorous compounds with cumyl hydroperoxide at 30 °C

	$10^3 . k$ (mol^{-1} s^{-1})
P[O–tert.Butyl]$_3$	220
P[O–⟨ ⟩]$_3$	31
(cyclic phosphite structure)	37

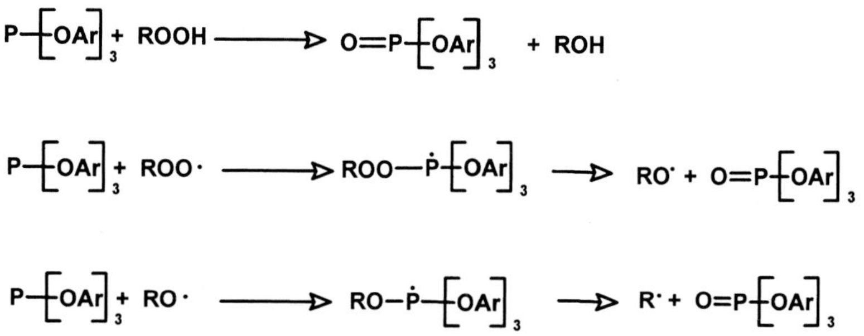

$$P\text{---}\left[OAr\right]_3 + ROOH \longrightarrow O{=}P\text{---}\left[OAr\right]_3 + ROH \tag{2.14}$$

$$P\text{---}\left[OAr\right]_3 + ROO\cdot \longrightarrow ROO\text{---}\overset{\cdot}{P}\text{---}\left[OAr\right]_3 \longrightarrow RO\cdot + O{=}P\text{---}\left[OAr\right]_3 \tag{2.15}$$

$$P\text{---}\left[OAr\right]_3 + RO\cdot \longrightarrow RO\text{---}\overset{\cdot}{P}\text{---}\left[OAr\right]_3 \longrightarrow R\cdot + O{=}P\text{---}\left[OAr\right]_3 \tag{2.16}$$

Scheme 2.16. Hydroperoxide decomposition and reactions of peroxy- and alkoxy-radicals in presence of aromatic phosphites

To what extent a trivalent phosphorus compound reacting with a peroxy radical, ROO$^\bullet$ contributes to the inhibition of autoxidation is still an open question. Reactions with the alkoxy radical, RO$^\bullet$, have also been postulated in the literature [127].

Because phosphites and phosphonites tend to hydrolyse, in practice mainly hydrolysis stable derivatives are being used. These are based generally on sterically hindered phenols. Because of their high reactivity, phosphites and phosphonites are used as stabilizers during processing in the melt (temperatures up to 300 °C). As "long term" stabilizers their contribution to the stabilization of the end product is small.

2.1.3.2
Organosulphur Compounds

Organosulphur compounds such as sulfides, dialkyl dithiocarbamates or thiodipropionates are well known hydroperoxide decomposers [128]. Their effect is based on the ability of sulfenic acids to decompose hydroperoxides. To this end, however, the sulfenic acid has to be formed by decomposition of the intermediate sulfoxides. The rates of decomposition of some sulfoxides are summarised in Table 2.5 [129].

Intermediates with β, β-sulfinyldipropionate structure (I) are particularly reactive with regard to sulfenic acid formation. For this reason, compounds such as di-stearyl, or di-lauryl dithiopropionate are mainly used (Scheme 2.17).

The sulfide reacts stoichiometrically in a first step with a hydroxy-peroxide molecule forming the "oxide". Sulfenic acid is formed through thermal decomposition. Further possible reactions are the formation of the sulfone according to Eq. (2.18) or oxidation with hydroperoxides, leading to sulfuric acid and other sulfur-containing oxidation products [130].

Table 2.5. Rate constants for the decomposition of sulfoxides at 100 °C

R^1 $\diagdown$ S → O R^2 $\diagup$		10^6 k (s^{-1})
R^1	R^2	
n - Propyl	n - Propyl	0.06
$CH_3CH_2\overset{O}{\overset{\|}{C}}OCH_2CH_2$—	$CH_3CH_2\overset{O}{\overset{\|}{C}}OCH_2CH_2$—	850

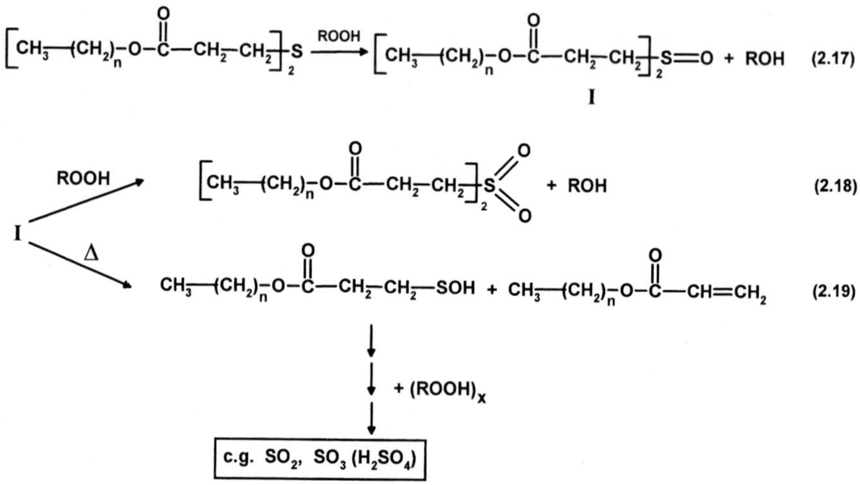

$$\left[CH_3\text{—}(CH_2)_n\text{—}O\text{—}\overset{O}{\overset{\|}{C}}\text{—}CH_2\text{—}CH_2\right]_2 S \xrightarrow{ROOH} \left[CH_3\text{—}(CH_2)_n\text{—}O\text{—}\overset{O}{\overset{\|}{C}}\text{—}CH_2\text{—}CH_2\right]_2 S{=}O + ROH \quad (2.17)$$

$$I$$

$$I \underset{\Delta}{\overset{ROOH}{\diagup\diagdown}}$$

$$\left[CH_3\text{—}(CH_2)_n\text{—}O\text{—}\overset{O}{\overset{\|}{C}}\text{—}CH_2\text{—}CH_2\right]_2 S\diagup\diagdown\overset{O}{\underset{O}{}} + ROH \quad (2.18)$$

$$CH_3\text{—}(CH_2)_n\text{—}O\text{—}\overset{O}{\overset{\|}{C}}\text{—}CH_2\text{—}CH_2\text{—}SOH + CH_3\text{—}(CH_2)_n\text{—}O\text{—}\overset{O}{\overset{\|}{C}}\text{—}CH{=}CH_2 \quad (2.19)$$

$$\downarrow + (ROOH)_x$$

$$\boxed{\text{c.g. } SO_2, \ SO_3 \ (H_2SO_4)}$$

Scheme 2.17. Hydroperoxide decomposition with thiodipropionate esters

These reaction products are catalysts for the decomposition of hydroperoxides. Because sulfoxides and the subsequent oxidation products are formed relatively slowly, thiosynergists primarily contribute to extension of the lifetime of plastics in the course of use at temperatures up to 150 °C. During processing in the melt, thiosynergists do not contribute to stabilization in contrast to trivalent phosphorus compounds [131].

Other known hydroperoxide decomposers are based on metal complexes of dialkyl dithiocarbamates [132].

2.1.4
Bifunctional Stabilizers

Bifunctional stabilizers are compounds having different functions combined in the same molecule. Particularly well investigated are sterically hindered phenols with sulphur substituents (Scheme 2.18).

These compounds behave as H-donors (chain breaking antioxidants) as well as hydroperoxide decomposers according to the described principles [133, 134].

Recently described in the literature [135] are bifunctional stabilisers based on molecular combinations of phosphites and phosphonites with sterically hindered amines. Such combinations are said to contribute to hydroperoxide decomposition, and also to thermal and photochemical stability.

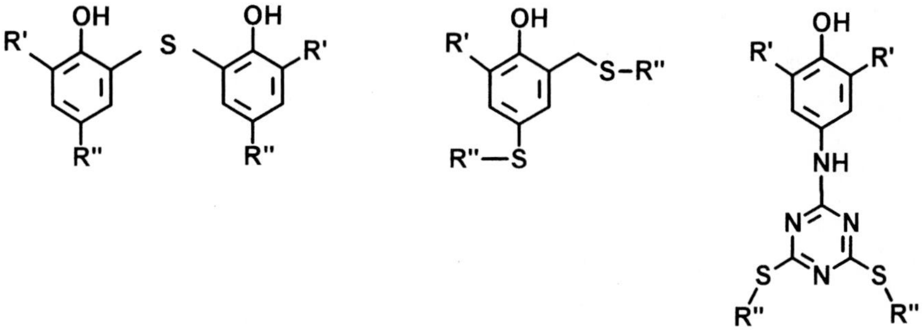

Scheme 2.18. Stabilizers with dual functionality

2.1.5
Blends of Stabilizers with Differing Functions

Physical blends of stabilizers with the same effective mechanism are referred to as homosynergistic combinations (e.g. phenols differing in steric hindrance).

Much more important are the so-called heterosynergistic combinations of stabilizers having different effective mechanisms, which complement each other.

Blends based on, e.g., the stabilizing effect of phenols or amines (CB-mechanism) and the hydroperoxide decomposition effect of phosphites and phosphonites [136] or thiosynergists [137] generally exhibit a "synergistic" effect (Fig. 2.2). Phenols contribute to the stabilization of the polymer during processing and long term thermal aging. The phosphites and phosphonites protect the polymer during processing,

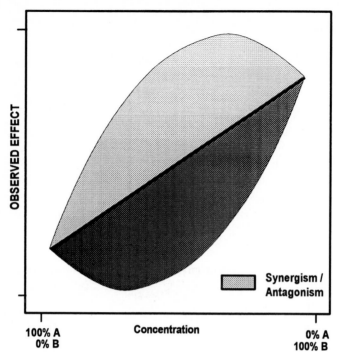

Fig. 2.2. Synergistic and antagonistic effects

while thiosynergists, in addition to the phenols, contribute to long term thermal stabilization.

The opposite, i.e. if the observed effect is less than the contribution of the individual components, is referred to as antagonism. Antagonistic effects have been observed occasionally in combinations of phenolic antioxidants and sterically hindered amines [138].

Recent investigations [139] have shown, however, that such combinations are quite able to exhibit synergistic effects in long term aging.

2.2
Stabilization Against Photooxidative Degradation

In contrast to thermooxidative degradation, photooxidative degradation of polymers, as the name implies, is initiated by the action of photons on the polymer.

The following autoxidation proceeds, however, analogous to the already described autoxidation chain of reactions. The possibilities available for the inhibition of photon induced degradation are shown in Fig. 2.3.

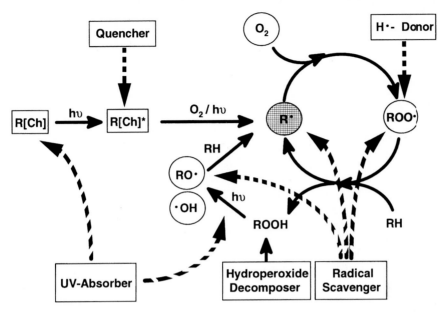

Fig. 2.3. General scheme of inhibition of photooxidation

By the use of suitable UV absorbers incorporated into the polymer, the penetrating light is absorbed and extremely rapidly deactivated by, e.g., transforming it to thermal energy by radiationless processes. These processes compete with the light-induced reactions of the polymer such as the photolysis of hydroperoxides, Norrish type I and type II reactions, or the photo-Fries reaction. The use of so-called "quenchers" deactivates the excited chromophores such as the carbonyl groups in polymers formed by thermooxidation. The latter, as shown, are efficient sensitizers for the photolysis of hydroperoxides. Finally, radical scavengers and hydroperoxide decomposers could be used in the same way as for the inhibition of thermally initiated autoxidation for the prevention of the chain reaction.

2.2.1
UV Absorbers

UV absorbers are colourless compounds characterized by high extinction coefficients in the spectral range of 300–400 nm. Dissolved in the polymer matrix, they absorb the UV portion of sunlight and compete with the substrate-specific absorption. By absorption of light, UV absorbers are transformed into an excited state, which, by rapid intramolecular processes, is deactivated and returned to its original state.

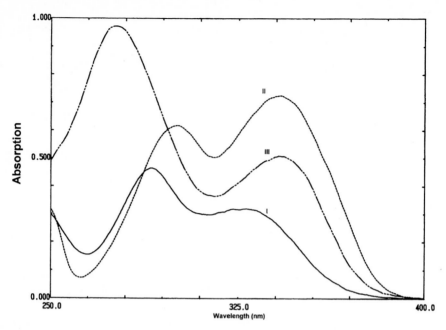

Fig. 2.4. UV-Absorption spectra of UV absorbers I=UVA-2, II=UVA-6, III=UVA-17 (see Appendix 3)

Consequently, the energy imposed on the polymer matrix cannot initiate the damaging, photooxidative reactions.

Currently used UV absorbers have an intramolecular hydrogen bond as in, e.g., o-hydroxybenzophenones, 2-(2-hydroxyphenyl)benzotriazole, 2-(2-hydroxyphenyl)-1,3,5-triazine, oxanilides, salicylates and cinnamates. Figure 2.4 shows the absorption spectra of three representatives of the above mentioned compound categories.

The efficiency of a UV screening agent in a non-light-absorbing substrate, e.g. PE-LD, PP, PMMA, depends on the concentration of the UV absorber used and the thickness of the polymer under consideration according to the law of Beer-Lambert as shown in Fig. 2.5 for UVA-6.

Energy dissipation can occur and is illustrated in Scheme 2.19, taking o-hydroxybenzophenone as an example.

In the first electronically excited singlet state, S_1 proton transfer takes place (excited state intramolecular proton transfer, ESIPT) forming the tautomeric "o-quinoide" keto form. Subsequently, an extremely rapid and radiationless deactivation of the S_1 state takes place reforming the molecule in its original state. For benzophenones it is postulated that deactivation of the tautomeric compound can also proceed

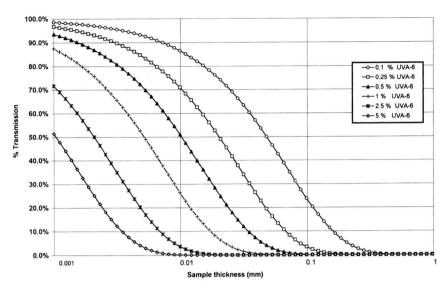

Fig. 2.5. Transmission of UV-light at 325 nm in relation to UV absorber concentration and sample thickness

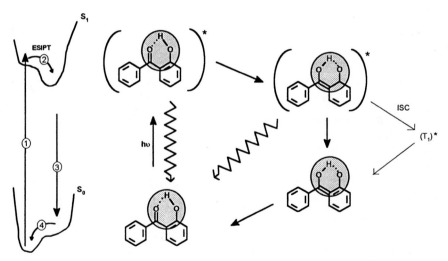

Scheme 2.19. Energy dissipation involving *o*-hydroxybenzophenones

by inter-system crossing (ISC) through the formation of the triplet state T_1.

Numerous reviews are published in the literature regarding the mechanism of the function of such UV-absorbers [140–144].

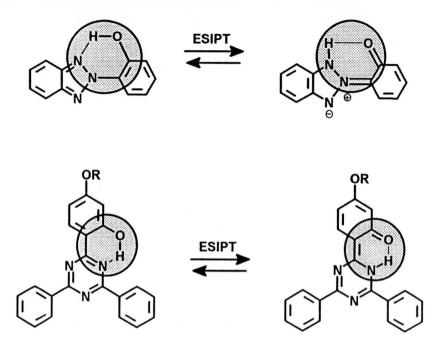

Scheme 2.20. Energy dissipation involving 2-(2-hydroxyphenyl)benzotriazoles or 2-(2-hydroxyphenyl)-1,3,5-triazines

Similar mechanisms are valid for energy dissipation and function of 2-(2-hydroxy-phenyl)benzotriazoles and 2-(2-hydroxyphenyl)-1,3,5-triazines (Scheme 2.20).

In the case of the 2-(2-hydroxyphenyl)benzotriazoles, the proton transfer in the excited singlet state is followed by very fast radiationless deactivation processes which are promoted by internal vibrations and rotations of the molecule [144, 145].

According to all these mechanisms, the absorbed energy is released as "harmless" heat to the environment. The UV absorber remains intact as long as intramolecular proton transfer can take place [146].

The lifetime of the mentioned UV absorbers in a given substrate decreases in general in the following order:
2-(2-hydroxyphenyl)-1,3,5-triazine > 2-(2-hydroxyphenyl)benzotriazole > 2-hydroxybenzophenone.

However, it also depends significantly on the other substituents on these molecules and the substrate in which they are dissolved. In their excited state all of these UV absorbers display "phenolic" structural elements. In principle, they are, therefore, capable of undergoing chemical reactions with radicals such as peroxy radicals, ROO• [147–

149]. Such reactions are possible if the intramolecular hydrogen bond is slightly disturbed [150] by, e.g., functional groups of a polar substrate or non-planar arrangement in the molecule. 2-Hydroxybenzophenones seem to contribute as antioxidants [151] at temperatures around 100 °C.

2.2.2
Quenchers

"Quenching" or "deactivation" of excited chromophores in polymers which, as shown, act as sensitizers for the photolysis of hydroperoxides, should contribute to the stabilization against light-induced degradation. In addition, quenching of the excited polymer-oxygen-exciplexes prevents or inhibits photooxidation of such polymers. This particularly concerns the photooxidation of polyethylene.

The most important and thoroughly investigated group are nickel chelates such as those shown in Scheme 2.21.

The absorption of such Ni-chelates in the region of 300–400 nm is small, but their contribution to light stability of the polymer can be significant [152]. Their function is based on an energy transfer of the excited chromophores to the quencher and subsequent deactivation of the excited quencher by radiation-free processes (Scheme 2.22).

Chien and Connor [153] have shown that in the photooxidation of cumene in the presence of diethyl ketones as sensitizers, such nickel chelates indeed effectively quench the triplet state of the ketone. Energy transfer can proceed by a "long range" mechanism as proposed by Foerster. The distance between chromophore and quencher can reach as high as 10 nm, providing there is significant overlap between the emission spectrum of the chromophore and the absorption spec-

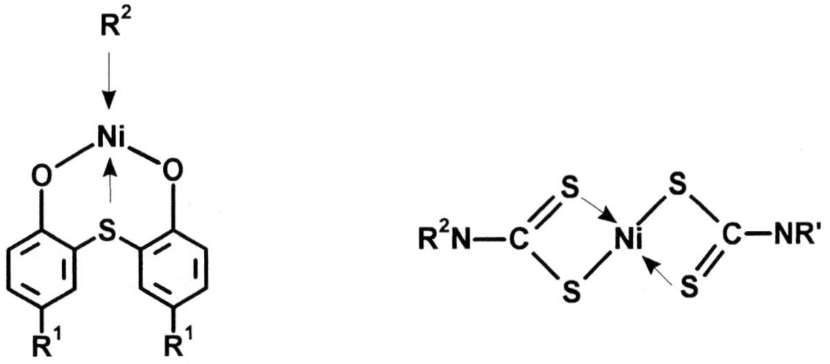

Scheme 2.21. Structure of some Ni-quenchers

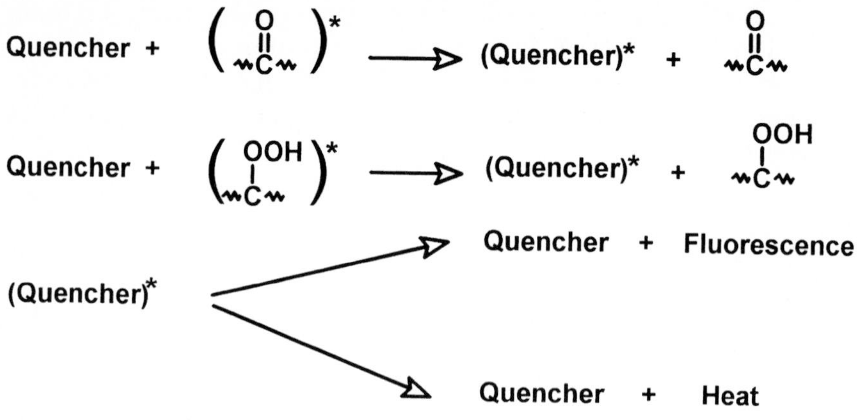

Scheme 2.22. Quenching of excited states

trum of the quencher. The other possibility is based on "collisional" energy transfer, whereby the distance between chromophore and quencher should not exceed 1.5 nm. Heller and Blattmann [145] arrived at the conclusion that, at the usual quencher concentrations of up to 1% relative to the polymer matrix, only quenching of excited singlet states over a dipole-dipole interaction contributes effectively to the stabilization.

Furthermore, it appears that most nickel quenchers contribute to hydroperoxide decomposition [154, 155] and also act as radical scavengers [156]. The effectiveness of the quencher is independent of the thickness of the polymer to be stabilized. For this reason such stabilizers are also effective regarding light stability in relatively thin films.

2.2.3
H-Donors and Radical Scavengers

2.2.3.1
Phenolic Antioxidants

Since the chain reaction of autoxidation proceeds in a similar way under thermooxidative and photooxidative conditions, the use of phenolic antioxidants should also contribute to stabilization under photooxidative conditions. However, the observed effect is relatively small [157] which is explained by the photochemically induced formation of phenoxyl radicals [158].

The photochemically induced formation of phenoxyl radicals competes with the stabilization of the peroxy radicals, ROO$^\bullet$ by transfer of

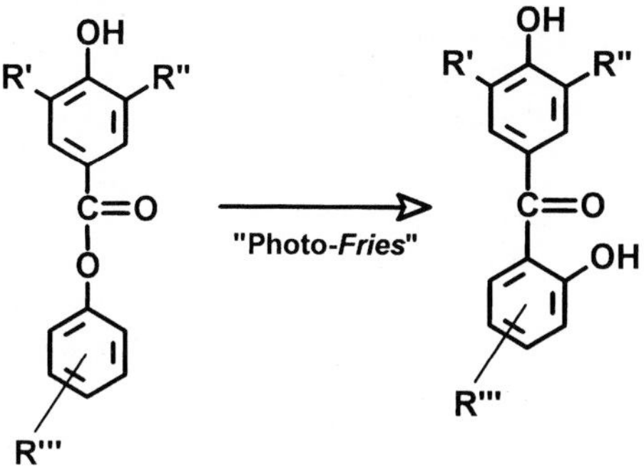

Scheme 2.23. Photo-Fries rearrangement of benzoate type phenol

hydrogen from the phenolic group resulting in the hydroperoxide, ROOH. Strengthening of the OH-bond by suitable substituents leads to improvement of the photo stability of sterically hindered phenols (Scheme 2.23).

In the above example, the corresponding o-hydroxybenzophenone is formed by a photo-Fries reaction and contributes further to light stabilization. Nevertheless, it is a fact that no sterically hindered phenol alone can effectively suppress photooxidation of polymer substrates.

2.2.3.2
Sterically Hindered Amines

Along with the preventive measures against light-induced degradation of polymers by the use of UV absorbers or quenchers, the application of radical scavengers contributes further to the stabilization and in many substrates decisively. In contrast to the sterically hindered phenols, the sterically hindered amines are extremely efficient stabilizers against the light-induced degradation of most polymers (hence, the original name: Hindered Amine Light Stabilizers, HALS, today frequently designated as Hindered Amine Stabilizers, HAS). Unlike UV-absorbers, and to a certain extent quenchers, sterically hindered amines do not absorb in the range 300–400 nm as Fig. 2.6 shows.

Their effectiveness against light-induced degradation of polymers, particularly of polyolefins, led to a revolution in stabilization.

The mechanistic aspects of their functioning are described in Schemes 2.10–2.12 and are also valid for the stabilization against photooxidation. A further significant contribution of HAS stabilizers is that they efficiently quench photooxidation initiated through a polymer/oxygen exciplex. A comprehensive study of the functioning of HAS was compiled recently by Gugumus [159].

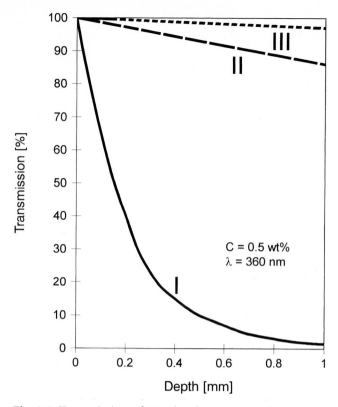

Fig. 2.6. Transmission of UV absorbers. I=UVA-2, II=Q-1, III=HAS-5 (see Appendix 3)

2.2.3.3
Hydroxylamines

Hydroxylamines appear to contribute to photo stabilization. The active species is probably the intermediate nitrone.

2.2.3.4
C-Radical Scavengers

So far there are no fundamental studies of the functioning of purely C-radical scavengers as stabilizers against photooxidation.

2.2.4
Hydroperoxide Decomposers

Hydroperoxide decomposers based on thiosynergists contribute in a way similar to that of interrupting the cycle of autoxidation as described in the example of thermooxidative inhibition. However, it should be pointed out that antagonistic effects may occur when thiosynergists and sterically hindered amines are used together.

2.2.5
Bifunctional Stabilizers

Molecular combinations of stabilizers with differing functional mechanisms for protection against photooxidation have not achieved the significance of the combinations described against thermooxidative degradation. Only the combination of a sterically hindered phenol with a HAS molecule is currently used against photooxidation of polyacetals and block copolymers having polyester and polyether segments (Scheme 2.24).

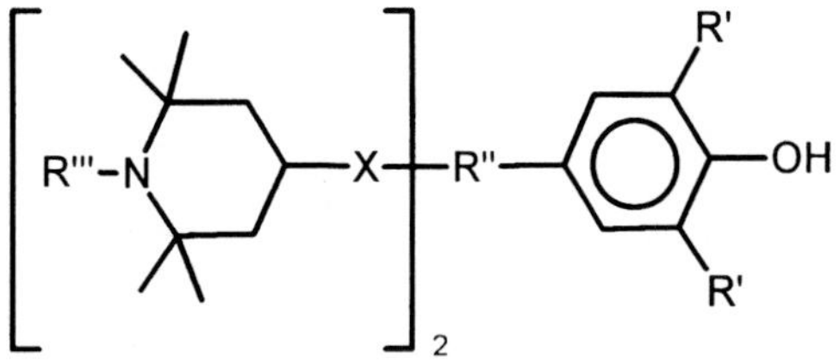

Scheme 2.24. Stabilizer with dual functionallity

2.2.6
Blends of Stabilizers with Differing Functional Mechanisms

Combinations of UV absorbers with various absorption spectra permit "covering" a broad range of sunlight's emission spectrum.

Combinations of benzotriazoles with oxanilides are characterized by high absorption in the short wavelength range of the UV spectrum. UV absorbers are deactivated, losing their absorption properties in the course of time, as shown, because of their partially phenolic character which can cause chemical reactions primarily by peroxy radicals. Combinations of UV absorbers with sterically hindered amines as scavengers of peroxy radicals generally exhibit enhanced effects. Combi-

Table 2.6. pKa values of sterically hindered amines in relation to their structures

	pKa	Ref.
$R-\overset{O}{\underset{\parallel}{C}}-O-\langle\text{piperidine}\rangle NH$	9.0	[161]
$R-\overset{O}{\underset{\parallel}{C}}-O-\langle\text{piperidine}\rangle N-CH_3$	8.9	[161]
$\left[O-\langle\text{piperidine}\rangle N-(CH_2)_2-O-\overset{O}{\underset{\parallel}{C}}-(CH_2)_2-\overset{O}{\underset{\parallel}{C}}\right]_n$	6.5	[161]
$R-(CH_2)_2-N\langle\text{piperidine}\rangle NH$	6.8	[162]
$R-\overset{O}{\underset{\parallel}{C}}-O-\langle\text{piperidine}\rangle N-O-\text{Alkyl}$	4.4	[164]

nations of sterically hindered amines with thiosynergists can exhibit antagonistic effects. Through hydroperoxide decomposition, thiosynergists form sulfuric acids in the course of transformation and form ammonium salts with sterically hindered amines. This ammonium salt formation dramatically reduces the photo stabilizing effect [160]. Similar negative effects are observed when halogen-containing flame retardants are used in combination with sterically hindered amines. Here too, halogen-containing acids are formed by light induced reactions, resulting in ammonium salt formation. Most recent work has led to the development of sterically hindered amines having low pKa value [161–164] (Table 2.6). Such compounds should be inert to deactivation by the above-mentioned salt formation mechanism.

Conclusions

Inhibition of thermooxidative and photooxidative autoxidation is possible to a large extent by the use of suitable stabilizers. Interrupting degradation by scavenging the reactive alkyl radicals under thermooxidative conditions is so far promising only under low oxygen concentration conditions, e.g. during processing of polymers in an extruder. With regard to peroxy radicals, antioxidants based on CB-acceptors as well as CB-donors are effective. In the presence of hydroperoxide decomposers, inert reaction products are finally formed. For the inhibition of light-induced degradation, UV absorbers, quenchers and C-radical scavengers such as sterically hindered amines can be used effectively.

This section describes only the principles of inhibition and the basic functional mechanisms. A prerequisite for the efficiency of stabilizers is their solubility in polymers and their availability in sufficient concentrations for inhibition. Furthermore, no physical loss of stabilizers should occur during the lifetime of a plastic article. The lifetime of the article is then determined only by the consumption of stabilizers caused by their function. The following section deals with the specific stabilization of the most important substrates. Solubility, diffusion and migration of stabilizers are dealt with in Chap. 5.

CHAPTER 3

Principles of Stabilization of Individual Substrates

This section is a summary of the application of the most common stabilizers protecting plastics against degradation during processing and protecting end products during their foreseen lifetime. The sequence of the review is by individual substrate. The additives used are listed in Appendix 3.

The effect of antioxidants and hydroperoxide decomposers on the stability of a plastic melt is determined by processing. Generally, the physical properties[1] of the melt are measured before and after processing, e.g. changes of the average molecular weight, e.g. viscosity of the melt, η_o or MFR[2], polydispersity of the polymer, M_w/M_n or melt volume-flow rate, MVR, and discoloration, yellowness index, Y.I.

Determination of the thermooxidative stability of a plastic part during actual use is carried out by aging at elevated temperatures in an air circulating oven. Subsequent to oven aging, mechanical values are determined such as elongation, impact strength, tensile strength, and also changes in average molecular weight, polydispersity of the polymer and discoloration.

The increase in carbonyl absorption in the infrared spectrum is also frequently determined as a criterion of oxidation. Oxygen uptake during aging is a further measure of oxidation sensitivity.

The use of thermoanalytical methods such as DSC measurements for the determination of aging processes is repeatedly proposed. However, this method is not suitable for predicting the long term lifetime of plastics parts. The various methods used are assessed in Chap. 6. Because the choice of test conditions influences the prediction of probable long term lifetime depending on the chosen stabilization system, careful assessment of the methods used and the selection of suitable tests is unavoidable.

---

[1] All standard testing methods used are summarized in Appendix 2
[2] All abbreviations used are summarized in Appendix 1

3.1
Polyolefins

3.1.1
Melt Stability of Polyolefins

The effect of sterically hindered phenolic antioxidants on the stability of a polyolefin under processing conditions in the melt is described extensively in the literature [27, 165]. In Fig. 3.1 is summarized the contribution of some phenols. For all structures of stabilizers see Appendix 3. All concentration are given in %, weight/weight! with different content of phenolic groups and different steric hindrance with regard to their effect in multiple extrusions of polypropylene (criterion: melt flow, MFR).

The effectiveness of a sterically hindered phenol depends, first, directly on the number of phenolic groups that contribute as H-donors, and second, on the steric hindrance in the 2- and 6-position [166]. Phenols with small steric hindrance in the 2- and 6-position exhibit in general better effectiveness than phenols having in these positions substituents with greater steric hindrance.

Best stability of polyolefins during processing is achieved with a combination of a sterically hindered phenol with a phosphite. Fig-

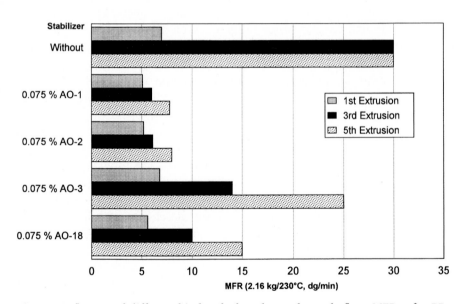

Fig. 3.1. Influence of different hindered phenols on the melt flow, MFR, of a PP-homopolymer during multiple extrusion at 270 °C

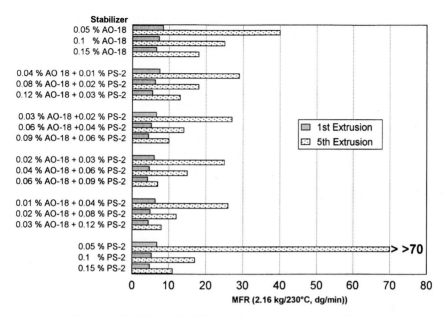

Fig. 3.2. Influence of a blend of a hindered phenol, AO-18, and a phosphite, PS-2, on the melt flow, MFR, of a PP-homopolymer during multiple extrusion at 280 °C

ure 3.2 depicts the results of multiple extrusions of polypropylene with phenol/phosphite blends having various antioxidant to phosphite ratios [136, 166].

It can be clearly seen that the combination phenol, AO-18/phosphite, PS-2, stabilizes the polymer melt more effectively than the individual components alone. Naturally, total concentration has great effect depending on processing temperature, shearing in the machine and further factors. Furthermore, stabilization depends on the choice of phenol and phosphite in the blend [167] and should be optimally adjusted for a given polymer.

Stabilization of polyethylene during processing in the melt is analogous to the described stabilization of polypropylene by using a blend of a sterically hindered phenol with a phosphite. This combination efficiently suppresses crosslinking of this PE-HD (Cr-catalyst) as shown in Fig. 3.3 [25, 27, 168].

In this instance too, effectiveness of the combination depends on the choice of the phenol and the phosphite as well as on the total concentration [136, 167].

In addition to stabilizers, acid acceptors, e.g. Ca-stearate, are added in the processing of polyolefins to bind residues of acids, e.g. from catalyst carriers such as $MgCl_2$ [169].

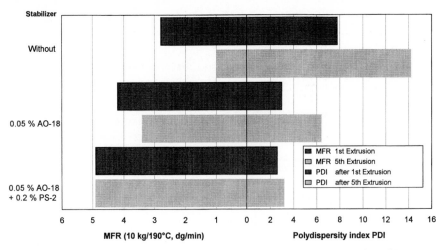

Fig. 3.3. Influence of a blend of a hindered phenol, AO-18, and a phosphite, PS-2, on the melt flow, MFR, and polydispersity, PDI, of a PE-HD (Cr-catalyst) during multiple extrusion at 260 °C

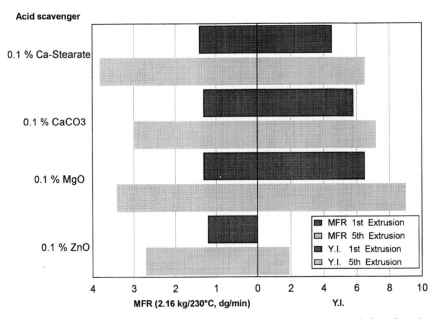

Fig. 3.4. Influence of acid scavengers on the melt flow, MFR and discoloration, Y.I., on processing of a PP-homopolymer at 280 °C (high yield catalyst technology). All samples contain 0.05% AO-18 + 0.1% PS-2

Such antacids can affect degradation of the polymer positively as well as negatively during processing and also its discoloration, as shown in Fig. 3.4.

The choice of the antacid depends on the polymer, the catalyst and its deactivation subsequent to polymerization, as well as on the phenol/phosphite combination, and has to be determined in some instances separately.

3.1.2
Long Term Thermal Stability of Polyolefins

Long term thermal stability of polyolefins, when using sterically hindered phenols as stabilizers of choice, is determined by the structure and concentration of the phenolic antioxidant. Figure 3.5 illustrates long term thermal stability of polypropylene stabilized with various phenols on aging at 135 and 149 °C.

AO-1 and AO-2 are too volatile at elevated aging temperatures, so that no stabilizing effect can be observed.

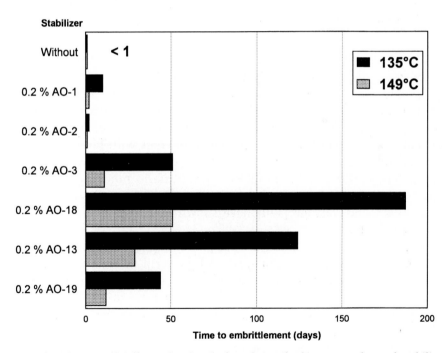

Fig. 3.5. Influence of different hindered phenols on the long term thermal stability (LTTS) of a PP-homopolymer at 135 and 149 °C. All samples contain: 0.1% Ca-stearate. Sample size: 1 mm compression molded plaques. Exposure device: circulating-air oven

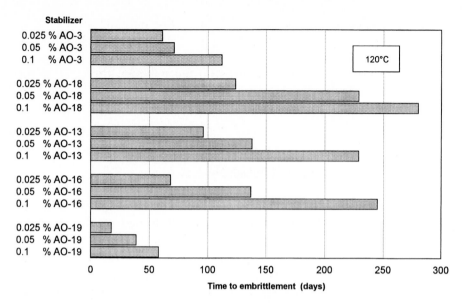

Fig. 3.6. Influence of various, hindered phenols on the long term thermal stability (LTTS) of a PE-HD (Cr-catalyst) at 120 °C. Sample size: 0.5 mm compression molded plaques. Exposure device: circulating-air oven

The results of long term thermal aging of high density polyethylene stabilized with various phenols are shown in Fig. 3.6

Phenols of the propionate type such as AO-3 and AO-18 exhibit greater effect compared with AO-19 because of their structure. This is to be expected on the basis of the reaction mechanism shown in Scheme 2.6. In long term thermal stress, phenols with high steric hindrance in the 2- and 6-positions exhibit better effectiveness at elevated aging temperatures than those with lower steric hindrance, in other words, the exact opposite of the situation in the stabilization of the melt. Glass and Valange [170] have shown that aging temperature influences results.

In contrast to stabilization during processing, the presence of phosphites has no significant influence on long term thermal aging of polyolefins. Zweifel and co-workers [168] have shown that the phosphite PS-2 is completely transformed after a short time to the corresponding phosphate and long term aging depends only on the concentration of the phenol used.

The addition of thiosynergists, however, imparts significant improvement of thermal long term stability of polymers stabilized with phenols as primary antioxidants. Figure 3.7 depicts a comparison of long term aging of polypropylene with and without thiosynergist [171].

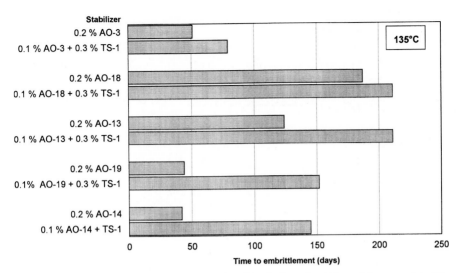

Fig. 3.7. Influence of blends of various, hindered phenols with a thiosynergist, TS-1, on the long term thermal stability (LTTS) of a PP-homopolymer at 135°C. All samples contain: 0.1% Ca-stearate. Sample size: 1 mm compression molded plaques. Exposure device: circulating-air oven

The contribution of thiosynergists is particularly high in combinations with sterically hindered phenols having inherently low effectiveness. Because thiosynergists can form oxidation products that have an organoleptic effect, in certain applications, e.g. in contact with potable water, undesirable effects may arise.

Recently it was realized that sterically hindered amines, HAS, based on tetramethyl piperidine compounds are good stabilizers with regard to long term thermal stability of polyolefins [172–176]. Figure 3.8 represents a summary of aging data obtained with polypropylene stabilized with various HAS. However, the polymers must be protected during processing, and the preferred combination is a phenol/phosphite combination in low concentrations.

Sterically hindered amines, HAS, also contribute significantly to long term thermal aging of polyethylene as shown in Fig. 3.9.

The low molecular weight HAS-5 is too volatile to contribute at these aging temperatures and test duration. Generally, the effectiveness of high molecular weight HAS in the stabilization of polyolefins increases with decreasing aging temperature [177]. Data obtained so far indicate the enormous potential of sterically hindered amines, HAS, with respect to long term stabilization. To avoid as far as possible loss of stabilizers by migration or evaporation from the surface of the

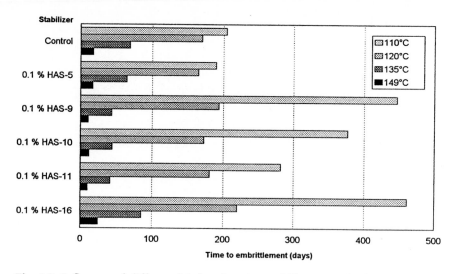

Fig. 3.8. Influence of different hindered amine stabilizers on the long term thermal stability (LTTS) of a PP-homopolymer at 110, 120, 135 and 149 °C. All samples contain: 0.05% AO-18 + 0.1% PS-2 and 0.05% Ca-stearate. Sample size: 1 mm compression molded plaques. Exposure device: circulating-air oven

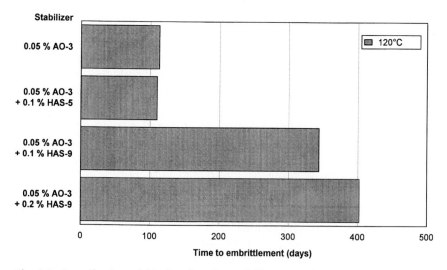

Fig. 3.9. Contribution of hindered amine stabilizers on the long term thermal stability (LTTS) of a PE-HD (Ti-catalyst) stabilized with a sterically hindered phenol, AO-3, at 120 °C. All samples contain: 0.1% Ca-Stearate. Sample size: 0.5 mm compression molded plaques. Exposure device: circulating-air oven

polymers, the products of choice are HAS derivatives with relatively high molecular weight or oligomers [172].

HAS stabilizers are protonated by acids and thus deactivated. Such acids could be formed, e.g., when sulfur and halogen-containing agrochemicals are used in greenhouses, or formed "in situ" when using halogen-containing flame retardants in the polymer.

In such instances, low or non-basic HAS derivatives such as NOR HAS compounds could be used (see Table 2.6).

3.1.3
Color Development in Polyolefins

The use of sterically hindered phenols as stabilizers may lead to discoloration of the substrate. Discolorations are generally due to transformation products with quinoid structure as a result of the way phe-

Table 3.1. Spectral data of some quinonemethides

Structure	Visible Absorption	
	$\lambda_{max.}$	ε $I. Mol^{-1}. cm^{-1}$
[chemical structure]	452	10^5
[chemical structure with $COOC_{18}H_{37}$ groups]	440	$3.5 \cdot 10^4$
[chemical structure with $COOC_{18}H_{37}$ groups]	420	10^2

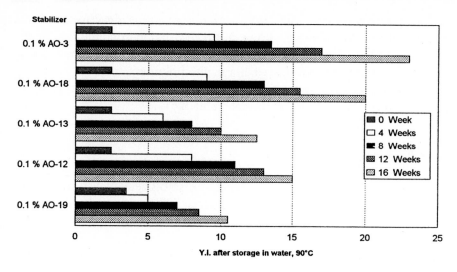

Fig. 3.10. Influence of various hindered phenol type stabilizers on the discoloration, Y.I. of PP-homopolymer after storage in water at 90 °C. All samples contain: 0.025% AO-3 + 0.1% PS-2, 0.05% Ca-stearate, 0.03% Dihydrotalcite, DHT-4A. Sample size: 1 mm compression molded plaques

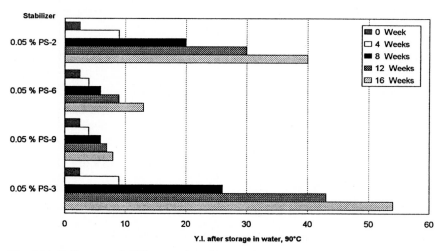

Fig. 3.11. Influence of different, phosphorous based processing stabilizers, on the color formation (Y.I.) of PP-homopolymer plaques after storage in water at 90 °C. All samples contain: 0.025% AO-3 + 0.05% AO-18, 0.05% Ca-stearate, 0.03% Dihydrotalcite, DHT-4A. Sample size: 1 mm compression molded plaques

nols react (see Scheme 2.7). Their generation depends on the substitution pattern of the phenol used. Discoloration of polyolefins is particularly pronounced if they are in continuous contact with water or, e.g., if exposed to NO_2 gases ("Gas Fading") [178]. Klemchuk and Horng [179] have determined the spectral properties of some quinone-methides. The results are summarized in Table 3.1.

The relationship between discolouration of PP homopolymer plaques after storage in water at 90 °C and the structure of the phenol used is shown in Fig. 3.10 [180].

The use of a suitable phenol/phosphite combination can significantly influence discoloration as shown in Fig. 3.11 [181]. An unequivocal explanation for this finding is not yet available.

3.1.4
Light Stability of Polyolefins

Stabilizers with different effectiveness principles can be used to inhibit light-induced degradation of polyolefins. UV absorbers act as "filters", quenchers deactivate excited states of chromophores in polymers by energy transfer, and suitable radical scavengers inhibit photo oxidation initiated by light (see Sect. 2.2).

Sunlight as well as artificial light sources can be used as radiation source. In the latter instance, irradiation can be carried out under dry, or wet, or alternating conditions, so that an acceleration compared with natural outdoor weathering is achieved. However, it has to be ensured that the irradiation equipment used yields results correlating with those obtained with outdoor weathering [182] (Sect 6.4).

Gugumus [183, 184] extensively investigated light stabilizers having different effectiveness principles with regard to their contribution to long term stabilization of polypropylene. Some results are listed in Fig. 3.12. Duration of irradiation was measured to retained tensile strength of 50%.

The results in Fig. 3.12 show that the low molecular weight sterically hindered amine, HAS-5, imparts the best protection against photooxidative degradation of PP-film strips. The phenolic antioxidant AO-28 also imparts a certain degree of light stabilization. In this instance, the sterically hindered phenol is protected by "in situ" formation of an o-hydroxybenzophenone group through a photo-Fries reaction.

These results show that polypropylene is generally better protected by radical scavengers than by UV absorbers. The use of nickel quenchers is increasingly being questioned because of eco-toxicological effects.

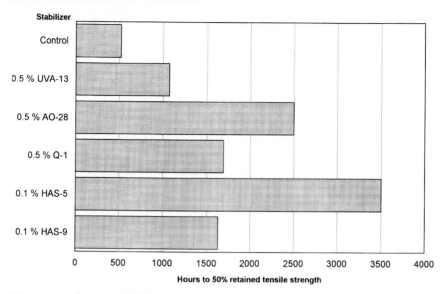

Fig. 3.12. Influence of different classes of light stabilizers on their performance in PP-homopolymer. All samples contain: 0.05% AO-18, 0.05% PS-2 and 0.1% Ca-stearate. Sample size: 50 μm extruded tapes, draw ratio 1:6. Exposure device: XENO 1200, b.p. temp. 55 °C, dry

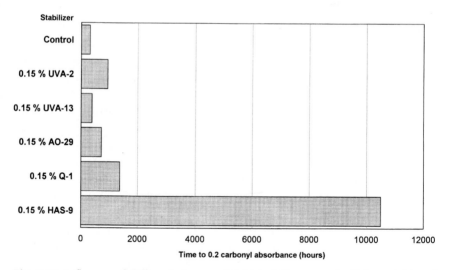

Fig. 3.13. Influence of different classes of light stabilizers on the light stability of a PE-LD-polymer. All samples contain: 0.03% AO-18. Sample size: 0.2 mm compression molded film. Exposure device: Weather.O.Meter, WRC 600, with water spraying

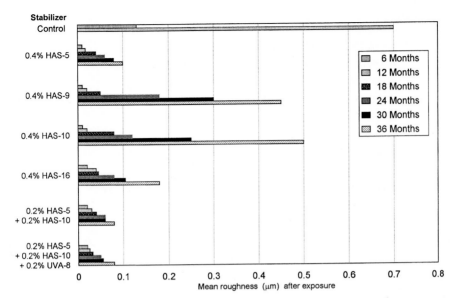

Fig. 3.14. Influence of the light stabilizer on the surface roughness of PP-homopolymer plaques after exposure in Florida. All samples contain: 0.05% AO-18, 0.1% PS-2 and 0.05% Ca-stearate. Sample size: 2 mm injection molded plaques. Exposure device: Florida, 45° south, 580 kJ cm^{-2} year^{-1}. Surface roughness: according to DIN 4762

Figure 3.13 summarizes the results relating to the stabilization of polyethylene against photooxidation by the use of various light stabilizers. Time of irradiation was measured to carbonyl absorption of 0.2 units. HAS-9 exhibits in this case the best effect.

Photooxidation is accelerated on the surface of the substrate because there irradiation energy and oxygen concentration are greatest.

Figures 3.14 and 3.15 summarize the results obtained from irradiation experiments with thick samples (2 mm) and a variety of HAS derivatives and combinations of HAS with different molecular weight [185]. HAS-5, a sterically hindered amine with low molecular weight, exhibits greater efficiency than HAS derivatives with high molecular weight. By migration from the sample's interior, the low molecular weight HAS-5 continuously protects the surface. Combinations of low and high molecular weight HAS are particularly effective. The use of UV absorbers can contribute further to the protection of the surface, but is generally necessary only for the protection of pigmented substrates.

The results show that best perfomance is achieved using combinations of low and high molecular weight HAS derivatives. Furthermore,

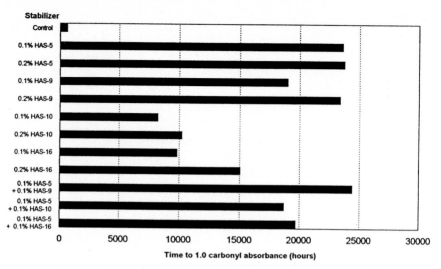

Fig. 3.15. Influence of the hindered amine stabilizer on the light stability of a PE-HD-homopolymer (Ti-catalyst). All samples contain: 0.03% AO-3 and 0.1% Ca-stearate. Sample size: 2 mm injection molded plaques. Exposure device: Weather.O.Meter, CI 65, b.p. temp. 63 °C, dry

long term thermal stability is ensured by the presence of high molecular weight HAS. Stabilization of polyolefins with sterically hindered amines in conjunction with only low phenol concentrations as basic stabilization also generally imparts excellent stability against discoloration in water storage or in contact with NO_2 gases [186]

3.2
Thermoplastic Polyolefins ("TPO")

The "classic" elastomer-modified polypropylene is prepared by compounding polypropylene with an elastomer of the EPR or EPDM type rubber. The development of special, high yield Ziegler-Natta catalysts made it possible to prepare elastomer-modified polypropylene by sequential copolymerization in the reactor.

Stabilization of these materials during processing and use under long term conditions such as heat and light is carried out according to the same principles as those for the stabilization of polyolefins. Combinations of stabilizers based on sterically hindered phenols such as AO-3 and AO-18 and phosphites as PS-2 are used for the stabilization during processing in the extruder and in injection moulding machines. In the course of compounding, during processing of the melt, substantial shear forces arise and thus the elastomer has to be protected by the addition of appropriate stabilizer concentrations. Meyer and Zweifel have shown [187, 188] that best stabilization of automotive bumpers is achieved by combinations of HAS compounds having low, and high molecular weight such as HAS-5/HAS-10 with regard to long term service life. Because impact-modified polypropylene represents a multi phase system, stabilizers are more soluble in the amorphous elastomer than in the partially crystalline polypropylene matrix. Generally, increase of the stabilizer concentration leads to extension of the lifetime of the plastic part.

Protection of the elastomer phase against thermal, and photooxidative degradation is particularly important, because its oxidation leads to loss of impact strength.

With regard to the behaviour of elastomer-modified polypropylene under outdoor weathering conditions, no difference was observed between compounded and sequentially polymerized materials. Polypropylene containing PE-HD along with the elastomer component exhibited improved weathering properties [187, 188]. The use of UV absorbers contributes further, particularly in pigmented TPO.

The possibility of producing syndiotactic polypropylene with enhanced heat deflection temperature by using metallocene catalysts will increase the application of TPO in future. So far, there are no results available concerning its stabilization.

3.3
Elastomers

Elastomers, e.g. rubber, are used for a variety of applications. Within the framework of this monograph only technical applications of elastomers as impact modifiers are discussed as already outlined with regard to impact-modified polypropylene. The most common elastomers are styrene-butadiene copolymers, SBR, polybutadiene, BR, and ethylene-propylene copolymers, EPR, as well as ethylene-propylene-diene copolymers, EPDM.

It has been shown that elastomers with butadiene content are prone to gel formation caused by crosslinking during processing. Addition of phenolic antioxidants such as AO-2 or AO-3, preferably in combination with phosphites like PS-1, results in the desired stability. Addition of thiosynergists contributes significantly to long term stability during thermal aging. The use of phenolic antioxidants having sulfur atoms in the substitution pattern ("dual functionality") such as AO-23 or AO-25 contributes further to improvement regarding stabilization of elastomers [189]. Aminic antioxidants are not generally used in these applications because they lead to discoloration and do not have legislative approval for use in contact with food.

Elastomers used as impact modifiers are generally not specifically protected against photooxidation because sterically hindered amines, HAS, or UV absorbers are added to polypropylene or polystyrene as continuous phase during processing to the end product.

3.4
Styrenic Polymers

3.4.1
Polystyrene

Polystyrene, especially crystal polystyrene, inherently exhibits a relatively high stability regarding degradation during processing in the melt and thermooxidative degradation. However, because waste from the manufacture of foamed and thermoformed articles such as expanded polystyrenes is available in large quantities, it is recycled for new processing. As a result, degradation occurs caused by chain scission. Degradation can be substantially retarded by the addition of phenolic antioxidants, e.g. 0.1% AO-3 [190]. Figure 3.16 depicts the change of weight average, M_w, and number average, M_n, molecular

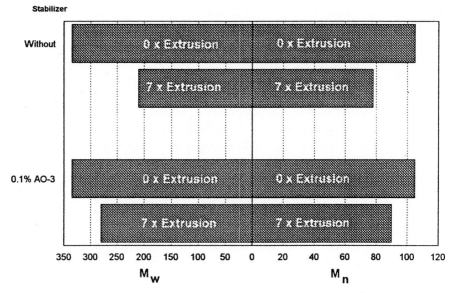

Fig. 3.16. Changes in M_w and M_n of polystyrene after multiple extrusions at 170 °C without stabilizer and in presence of 0.1% AO-3

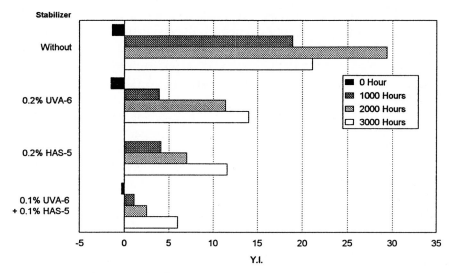

Fig. 3.17. Discoloration of polystyrene upon exposure to light. Sample size: 2 mm injection molded plaques. Exposure device: Weather.O.Meter WRC 600, dry , b.p. temp. 55 °C

weight after multiple extrusion at 170 °C of unstabilized material and in the presence of AO-3 added during polymerization [191]

Expanded polystyrene with a molecular weight of 280 000 exhibits good material properties, but polystyrene with less than 280 000 molecular weight cannot be foamed or thermoformed.

Addition of the phenolic antioxidant during polymerization leads to better results than subsequent addition prior to processing [191].

Phosphites and phosphite/phenol combinations generally suppress discoloration.

For the inhibition of photooxidative degradation, UV absorbers of the benzotriazole type are being used, or preferably, combinations of UV absorbers with sterically hindered amines, HAS. Figure 3.17 illustrates the connection between stabilization and discoloration, Y.I., after light exposure of polystyrene.

3.4.2
Impact-Modified Polystyrene

Thermooxidative degradation of polystyrene impact-modified with polybutadiene manifests itself in yellowing and loss of mechanical properties. Similar to EPR impact-modified polypropylene, the elastomer phase is oxidatively degraded.

The addition of antioxidants such as AO-3 and AO-10 effectively inhibits degradation. For homogeneous dispersion of the antioxidants, they are added during polymerization. It follows that they should have no effect on polymerization kinetics. Incorporation of the antioxidants during polymerization also partially inhibits crosslinking of the polybutadiene.

Styrene-butadiene, SB, and styrene-butadiene-styrene block copolymers, SBS, can be processed as transparent, impact-modified thermoplastics. It was found that the addition of a C-radical scavenger, AO-30 in combination with phenolic antioxidants like AO-3 and phosphites like PS-1 inhibits crosslinking of the butadiene phase [192].

For the inhibition of photooxidative degradation and in analogy with the stabilization of polystyrene, the same combinations of UV absorbers and sterically hindered amines, HAS, are used (Fig. 3.17).

3.4.3
Styrene/Acrylonitrile Copolymers

Styrene-acrylonitrile copolymers, SAN, become discolored under thermal stress above 220 °C, e.g. in injection moulding, even in the absence of oxygen. This is caused mainly by reactions of the acrylonitrile comonomer. The combination of a phenolic antioxidant such as AO-3 with a phosphite, PS-2 suppresses discoloration during processing [190].

The same criteria are valid against photooxidative degradation as for polystyrene and impact-modified polystyrene. Best results are achieved with combinations of UV absorbers and sterically hindered amines, HAS.

3.4.4
Acrylonitrile/Butadiene/Styrene Copolymers

Acrylonitrile-butadiene-styrene copolymers, ABS, are, from a technical point of view, the most important of all styrene copolymers. This polymer is the basic material for high value articles. Again, the elastomer phase is the oxidation-sensitive component in the copolymer.

ABS is manufactured by two different methods requiring different stabilization.

In the emulsion process, elastomer and rigid phase are prepared in separate steps. For the subsequent compounding the ratio of elastomer to rigid phase can be varied at will within certain limits. The production of the elastomer by grafting styrene and acrylonitrile onto the

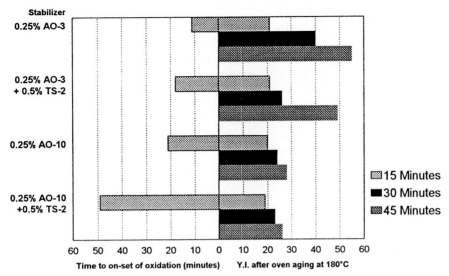

Fig. 3.18. Influence of antioxidants on the thermal stability of ABS graft-phase powder containing 25% polybutadiene. DTA: isothermal, 180 °C, air. Y.I.: oven aging at 180 °C

polybutadiene in latex form requires addition of the stabilizers to the emulsion in a suitable form prior to coagulation upon completion of the reaction. The subsequent drying of the product in powder form requires stabilizers that efficiently inhibit possible thermooxidation. Figure 3.18 presents a summary of data illustrating the influence of various antioxidants on the thermooxidative stability of the elastomer phase in ABS. Criteria were time to onset of exotherm determined by DTA and discoloration after aging in circulating air oven [191].

If a phosphite, e.g. PS-1, is added to the phenol/thiosynergist combination, additional improvement of oxidation inhibition can be achieved. In any event, this type of stabilization prevents spontaneous ignition of the powder caused by oxidation during drying.

If additional amounts of antioxidants should be needed for processing, they can be added to the dry powder prior to compounding.

In the manufacture of ABS by the mass polymerization method, polybutadiene is dissolved in the acrylonitrile-styrene monomer mixture. Grafting onto the polybutadiene phase and preparation of the SAN rigid phase proceed simultaneously. Isolation and drying of the oxidation sensitive ABS graft phase becomes redundant. The antioxidants AO-3, or preferably AO-10, in combination with thiosynergists can be added prior to granulation.

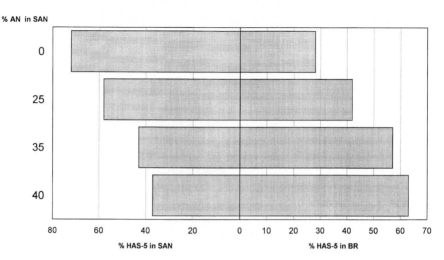

Fig. 3.19. Influence of acrylonitrile, AN, to styrene ratio in SAN phase on the percent of HAS-5 that partitions into BR phase [194]

The lifetime of ABS end products, analogous to impact-modified polystyrene, is determined by the oxidation of the elastomer particles. Hirai [193] found a correlation between thermooxidative stability of ABS and its content of polybutadiene phase. Kulich and coworkers [194] measured the partitioning of some stabilizers in both phases, BR phase and SAN phase. Figure 3.19 depicts partitioning of the sterically hindered amine, HAS-5, in the BR, and in the SAN phase as function of acrylonitrile content in the thermoplastic phase.

With increasing acrylonitrile content in the SAN phase, the amount of HAS-5 in the BR phase clearly increases. In this way, the desired protective effect of the oxidation sensitive elastomer phase is ensured. Kulich and co-workers [194] have furthermore shown that partitioning of stabilizers, e.g. thiosynergists, in multiphase systems correlates with solubility in the individual phases.

Stabilization of ABS against photooxidative degradation is preferably effected, as generally valid for styrenic polymers, with a combination of UV absorbers such as UVA-6 with a sterically hindered amine, HAS-5. However, ABS materials stabilized with this combination may exhibit yellowing after light irradiation and storage in the dark. Combinations of UVA-6 with HAS-16 or combinations of UVA-10 with HAS-6 do not lead to yellowing after storage in the dark [195].

3.5
Polyesters

Linear polyesters such as poly (ethylene terephthalate), PET, or poly (butylene terephthalate), PBT, are stabilized during processing in the melt using organo phosphorus compounds. The effectiveness of these compounds can be explained by their ability to form complexes with transesterification catalysts remaining in the polycondensate, e.g. organic compounds based on manganese, zinc or tin.

The addition of a phosphate like AO-8 at the beginning of the polycondensation of PET, which does not influence the kinetics of the condensation reaction, leads to a polymer with few carboxyl terminal groups and thus improves properties like hydrolytic stability of fibres and largely inhibits discoloration of the substrate after the condensation reaction [196, 197].

Degradation of polyesters during processing in the melt, e.g. extrusion or injection moulding, can be inhibited by the addition of combinations of phenolic antioxidants and phosphites. Figure 3.20 illustrates the change of the average molecular weight, M_n of PET in the course of multiple extrusions at 265 °C.

The effectiveness of phenolic antioxidants, particularly in combination with phosphites and phosphonites, manifests itself also in the

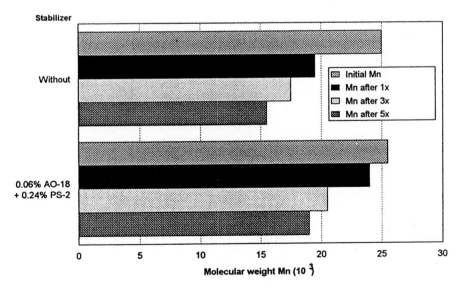

Fig. 3.20. Influence of a blend of a hindered phenol, AO-18, and a phosphite, PS-2, on the processing stability of a PET during multiple extrusion at 265 °C

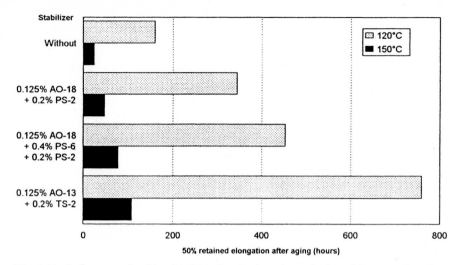

Fig. 3.21. Influence of a blend of hindered phenols and phosphites or phosphonites on the long term thermal stability (LTTS) of a PBT after oven aging at 120 and 150 °C. Sample size: 1 mm thick, injection molded dumb bells. Exposure device: circulating-air oven

maintenance of the mechanical properties of polyesters under long term thermal stress. Figure 3.21 depicts the results obtained from long term thermal experiments with PBT samples at 120 and 150 °C.

The contribution of a hydroperoxide decomposer like TS-2 towards stabilization during long term thermal aging is particularly notable. Sterically hindered amines, HAS, do not contribute towards stabilization of polyesters during long term thermal aging.

Discoloration of polyesters caused by long term thermal stress can also be avoided to some extent by the addition of phenol/phosphite combinations. Figure 3.22 presents a summary of the results obtained with regard to discoloration of PET after 1000 h aging as function of temperature [196, 197].

Poly (ethylene terephthalate) and poly (butylene terephthalate) exhibit better stability regarding photooxidative degradation than aliphatic polyamides and polyurethanes.

Because degradation occurs primarily on the surface, UV absorbers contribute to the light stability of polyester products. Figure 3.23 depicts results from light irradiation of PET fibres as function of stabilization formulations. The measured criterion was irradiation time to 20% loss of initial tensile strength at break.

The best effect was achieved with the combination of the UV absorber with the sterically hindered amine, HAS-5.

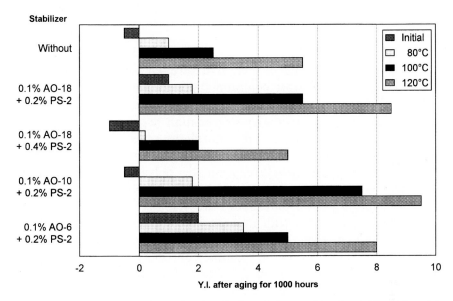

Fig. 3.22. Influence of different hindered phenols in combination with a phosphite, PS-2, on the discoloration, Y.I., of PET at 80, 100 and 120 °C. Sample size: 0.4 mm extruded ribbons. Exposure device: circulating-air oven

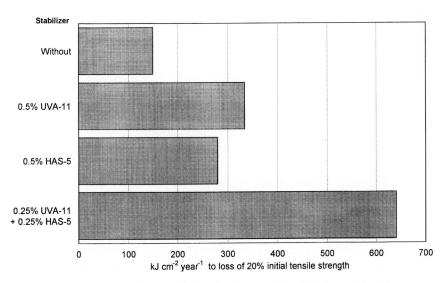

Fig. 3.23. Influence of UV absorbers and UV absorber/HAS combinations on the light stability of PET fibres. Sample size: fibres, knitted fabric. Exposure: Florida, 45° south, 580 kJ cm^{-2} year^{-1}, under glass

3.6
Polyamides

Aliphatic polyamides differ in their structure, PA 6,6, PA 6, PA 4,6, PA 11 and PA 12 are the most common types. Gijsman and co-workers [198] have shown, taking the stability of polyamide 6,6 and polyamide 4,6 as an example, that oxidative degradation depends on the crystallinity and on the density of the amorphous phase. This is because these two factors influence diffusion of oxygen into the polymer matrix. Because of their structure, aromatic polyamides are generally very stable with regard to thermooxidative degradation.

Traditionally, aliphatic polyamides are stabilized with copper salts. In some instances blends with manganese compounds are being used. Even small amounts of copper (up to 50 ppm) in combination with halogen ions such as iodine or bromine contribute significantly to the thermal stability of polyamides. The effect of copper salts as stabilizers is actually surprising because such copper ions act as prodegradants in polyolefins (Sect. 4.1).

The mechanism of stabilization with such copper/halogen salt combinations is still subject to scientific investigations. According to Janssen and co-workers [199] it may be due to decomposition of hydroperoxides initiated by metal ions (Chap. 1, Eqs. (1.16) and (1.17)) and further catalyzed by halogen-containing salts. Copper is added to the polymer as copper acetate and the halogenides as copper iodide (or bromide). Good dispersability of such salts is critical. Leaching of the salts, particularly of the halogenides, in contact with water or water/solvent mixtures may cause problems. Furthermore, the substrate could discolor. Consequently, such stabilizers are used, preferentially in carbon black and glass fibre filled applications.

The use of aromatic amines, which are actually good stabilizers in long term thermal applications, leads to discoloration. They are not suitable for applications requiring approval for contact with food for toxicological reasons.

Phenolic antioxidants contribute to the stabilization of aliphatic polyamides by improving initial colour after polycondensation and processing. Phenolic antioxidants are added to the condensation mass preferably prior to the termination of the polymerization reaction. The concentration of chain terminators has to be adapted corresponding to the desired product viscosity to compensate possible influence of the phenol.

The polymer melt's stability during processing does not generally present problems and the presence of phenolic antioxidants inhibits

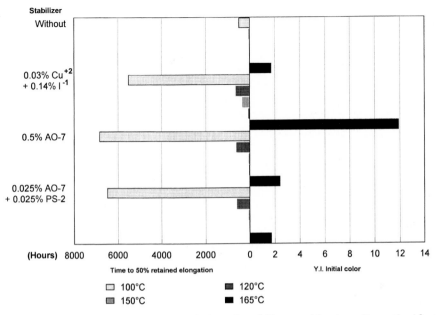

Fig. 3.24. Comparison of copper salt based stabilizers with phenolic antioxidant stabilizerson long term thermal stability (LTTS) and discoloration, Y.I., of PA 6 at various temperatures. Sample size: injection molded dumb-bells, 1 mm thick. Exposure device: circulating-air oven

possible discoloration after extrusion. Caution is indicated in the use of phosphites as process stabilizers under critical processing conditions such as the manufacture of thin-walled injection moulded articles, because the melt flow, MFR, may be decreased.

Figure 3.24 shows the results of long term aging experiments with PA 6 samples stabilized with copper salts and the phenolic antioxidant AO-7, as well as a combination of AO-7 with the phosphite, PS-2 [200, 201]. It was found that the phenolic antioxidant or its combination with the processing stabilizer is more effective than the stabilization with copper salt at low temperatures.

Moreover, initial colour prior to thermal aging is substantially better with the phenolic antioxidant or its combination, than with copper salt stabilization.

Figure 3.25 illustrates the results of thermal long term aging of PA-6, carried out with various phenolic antioxidants.

The exceptional effectiveness of AO-7 compared with the other phenolic antioxidants is probably due to improved compatibility with the polymer matrix as a result of the structural analogy of this AO with it.

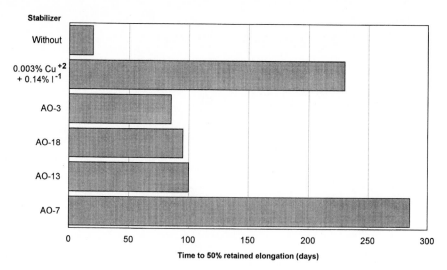

Fig. 3.25. Comparison of different hindered phenols and a copper iodide based stabilizer on the long term thermal stability (LTTS) of PA 6 after aging at 100 °C. Sample size: injection moulded dumb-bells, 1 mm thick. Exposure device: circulating-air oven

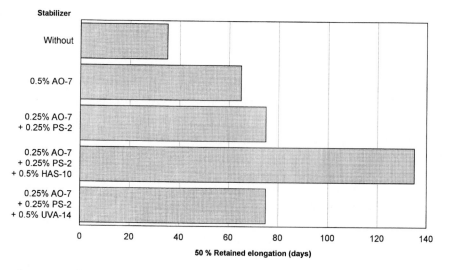

Fig. 3.26. Influence of different light stabilizers on the light stability of PA 6 fibres. Sample size: fibres, 110/24 dtex. Exposure device: Weather.O.Meter CI 65, b.p. temp. 65 °C, r.h. 60%, wet cycle 102/18 min

Stabilization of aliphatic polyamides against photooxidative degradation is very important. The use of stabilizers based on copper/halogen salts contributes significantly to stability. However, because of the

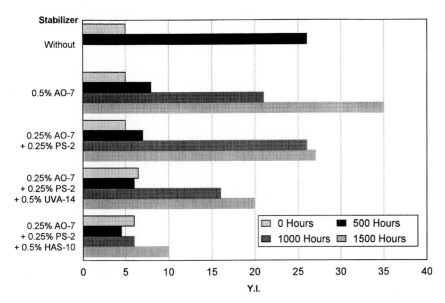

Fig. 3.27. Influence of different light stabilizers on the discoloration of polyamide 6 fibres upon irradiation. Sample size: fibres, 110/24 dtex. Exposure device: Weather.O.Meter CI 35, b.p. temp. 90 °C, r.h. 30%, according to FAKRA, DIN 75202, 1 cycle 96 h

halogen salts' extractibility, applications are limited [202]. The use of suitable UV absorbers and especially sterically hindered amines, HAS, contributes to the light stability. Figure 3.26 depicts the results of irradiation experiments with PA 6 fibres containing various stabilizers.

Figure 3.27 depicts the results of experiments related to discoloration of PA 6 fibres on irradiation.

The best results with respect to stabilization against photooxidative degradation are obtained with HAS-10 in combination with a phenol/phosphite combination as basic stabilization. In contrast to the possibility of adding the phenolic antioxidant during or towards the end of the condensation reaction, the addition of phosphites, HAS and UV absorber generally should not be carried out during the reaction.

Botkin and co-workers [202] have shown that the effectiveness of copper/halogen salt stabilizers can be improved further by the addition of HAS-10. Manganese salts bring about further improvement of stability in TiO$_2$ filled polyamides [203].

UV absorbers also contribute to improved stability, but are primarily used in pigmented materials. These findings apply generally to all aliphatic PA substrates in a variety of applications. A summary of further results, particularly concerning injection moulded articles, was given by Gugumus [204].

3.7
Polyacetals

Polyacetals are inherently unstable polymers because they degrade by depolymerization starting at the chain end. Through autoxidation, formic acid is formed causing acidolysis of the polymer thus accelerating degradation. Depolymerization of the homopolymer can be prevented by ether- or ester endcapping of the terminal hydroxyl groups. By the inclusion of suitable comonomers the alternating -C-O-C- structure is interrupted and the zipper-like depolymerization can be stopped.

To prevent acidolysis of the polymer, polyacetals and corresponding copolymers are protected by the addition of acid acceptors (Sect. 1.4.5).

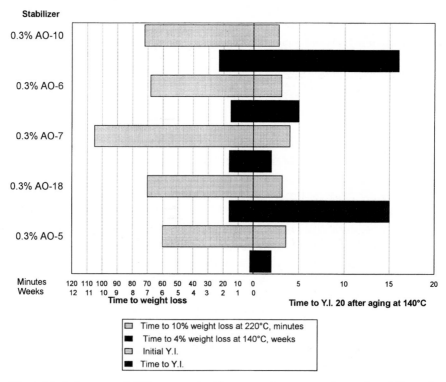

Fig. 3.28. Influence of different antioxidants on long term thermal stability of a Polyacetal-copolymer after aging at 220 and 140 °C. All samples contain: 0.3% Ca-stearate. Sample size: 2 mm injection molded plaques. Exposure device: circulating-air oven

The thermooxidative degradation of polyacetals and its copolymers leads to the formation of volatile degradation products along with weight loss and discoloration of the polymer [205]. Figure 3.28 depicts these effects as functions of the individual phenolic antioxidants used (calcium stearate was used as acid acceptor).

Time to weight loss of 10% measured by TGA at 220 °C indicates the polymer's stability under processing conditions, while measurement to a weight loss of 4% at 140 °C indicates stability under long term thermal stress. Time to reaching a yellowness index, Y.I., of 20 after aging at 140 °C indicates tendency to discoloration.

Best results are obtained with AO-10 which, because of its molecular structure, should be well compatible with the polymer.

The great contribution of AO-7 to stabilization at elevated temperature can be explained with its additional effect as an acid acceptor.

The use of phosphites as hydroperoxide decomposers does not contribute further to stability. This is explained by the rapid oxidation of formaldehyde by peroxides to formic acid [206]. Furthermore, it should be borne in mind that phosphites and their reaction products, depending on molecular structure, hydrolyze in the presence of moisture, forming acids.

Polyoxymethylene and the corresponding copolymers cannot be used in outdoor applications without adequate light stabilization. Sta-

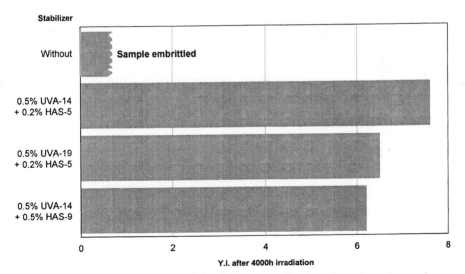

Fig. 3.29. Influence of different light stabilizers on discoloration of a polyacetal copolymer. All sample contain: 0.3% AO-9 + 0.3% Ca-stearate. Sample size: 1 mm compression molded plaques. Exposure device: XENO 1200, b.p. temp. 55 °C

bilization against photooxidative degradation is preferably achieved with a blend of UV absorbers and sterically hindered amines, HAS.

Figure 3.29 shows the results of irradiation experiments with a copolymer as a function of such combinations.

Because photooxidation begins on the polymers surface leading rapidly to crack formation and discoloration, the best protection is achieved with the combination of UVA-14 with the low molecular weight HAS-5, which easily migrates to the polymers surface.

3.8
Polycarbonates

Thermooxidative degradation of polycarbonate leads to discoloration of the clear, transparent substrate. Stabilization of polycarbonate should recognize that this polymer exhibits a strong tendency to hydrolyze in the presence of humidity. Pryde and Hellmann [207] have shown that hydrolysis depends on the number of phenolic terminal groups in the polymer. In using phosphites to suppress discoloration during processing or long term thermal aging, it should be ensured that no acid traces are present. The use of acid acceptors, like epoxides in combination with phosphites, or phosphonites, produces positive effects. Because processing temperatures are relatively high, stabilizers have to be thermally stable and non-volatile. Figure 3.30 depicts the change of melt viscosity of a polycarbonate with original melt

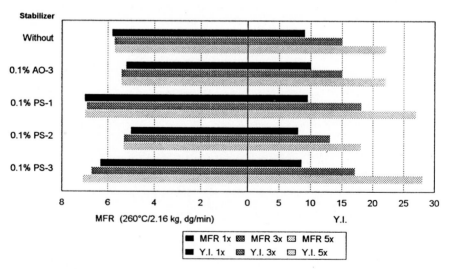

Fig. 3.30. Influence of different stabilizers on the melt flow, MFR, and discoloration, Y.I., of a polycarbonate after multiple extrusions at 330 °C

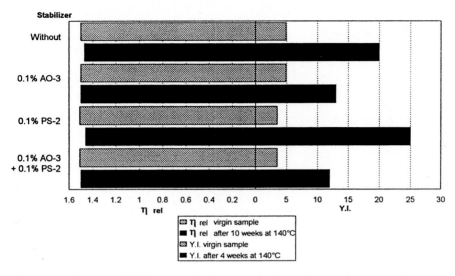

Fig. 3.31. Influence of different stabilizers on the long term thermal stability, η_{rel} and discoloration Y.I., of polycarbonate at 140 °C. Sample size: 2 mm injection molded plaques. Exposure device: circulating-air oven. η_{rel} in CH_2Cl_2

flow, MFR, of 5.0 (260 °C/2.16 kg, dg/min) as well as discoloration, Y.I., after multiple extrusions at 330 °C.

The phosphites PS-1 and PS-3 which tend to hydrolyze exhibit clearly weaker effect in the stabilization of the melt than the hydrolysis stable PS-2. Further, discoloration is best inhibited with the phosphite, PS-2, and the phenolic antioxidant, AO-3.

Figure 3.31 represents a summary of the results, change in molecular weight and discoloration, obtained in the stabilization of a polycarbonate under long term stress.

The combination AO-3/PS-2 exhibits the best contribution to thermal long term stability of this polycarbonate. However, it can be assumed that stability and tendency to discoloration of the substrate depends much on its quality, e.g. free phenolic groups.

Stabilization of polycarbonate against light induced degradation is extremely important. The originally ductile polycarbonate becomes brittle [208] under outdoor weathering conditions and degradation on the surface of the polymer sets in. Light stabilizers to be used must not negatively influence processability, i.e. decrease melt viscosity by molecular weight degradation, and long term thermal aging. The use of sterically hindered amines, HAS, is excluded because bases dramatically accelerate hydrolysis of polycarbonate. For this reason, UV absorbers are used for stabilization. They can be added directly to the poly-

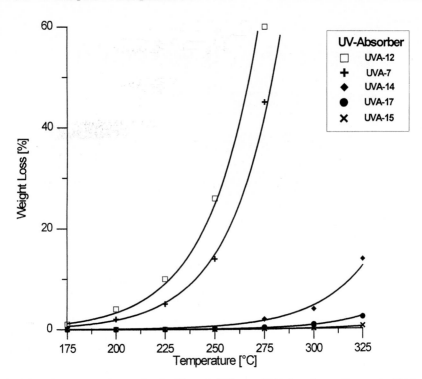

Fig. 3.32. Comparison of volatility of different UV absorbers. Dynamic TGA measurement, heating rate 20 °C/min, in air

mer prior to processing. Improved protection can be achieved by applying the UV absorber in high concentration onto the polymer's surface. The application is effected by coextrusion of a thin film using the same substrate or other film-forming polymers like poly (methylmethacrylate) along with the UV absorber [209, 210] onto the polymer's surface, by impregnation of the polymer's surface with a suitable UV absorber solution [211] or by varnishing of the polymer's surface with a varnish containing UV absorber [212].

The most elegant process is coextrusion of a thin film containing an efficient UV absorber. Because polycarbonate is processed at elevated temperatures, the UV absorbers of choice have to have as low volatility as possible. Olson and Webb [213] investigated the loss of various UV absorbers from a polycarbonate matrix below and above the polymer glass transition temperature. Only stabilizers with a molar volume of less than 230 cm^3/mol are lost at 195 °C through diffusion or evaporation.

Figure 3.32 shows volatility of some UV absorbers as a function of temperature [214].

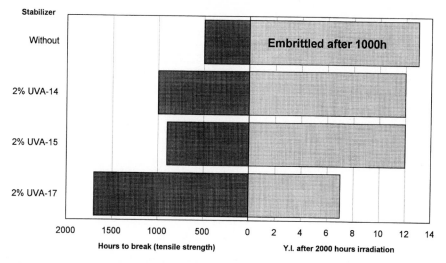

Fig. 3.33. Influence of UV absorbers on time to loss of tensile strength at break and discoloration, Y.I., of polycarbonate after exposure to light. Sample size: 20 μm films. Exposure device: Weather.O.Meter CI 65, b.p. temp. 63 °C, r.h. 60%, dry

Figure 3.33 represents a summary of the results from irradiation experiments with polycarbonate films stabilized with various UV absorbers. The parameter measured was irradiation time to loss of tensile strength at break and discoloration after irradiation.

At high UV absorber concentrations in the substrate, care should be taken that the additive does not precipiate on the surface (blooming). Further it is important that the polymer's heat deflection temperature is not decreased too much.

3.9
Polyurethane

Oxidative stability of polyurethanes depends on their structure, i.e. the chemical structure of the polyol, e.g. polyester, or polyether chains, and the isocyanate component. Thus, thermooxidative stability of polyurethanes with polyester segments is generally superior to that of polyether groups. Polyurethanes are commonly used as flexible foams, PUR foam, rigid microcellular foams, RIM-PUR, and thermoplastic polyurethane, T-PUR.

In the manufacture of PUR foams, a great amount of heat is released by the reaction of the polyol with the isocyanate (Sect. 1.4.7). Because of their structure, polyols tend to transform methylene groups vicinal to -O- groups to peroxide groups under thermooxida-

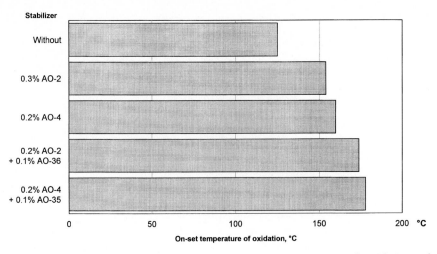

Fig. 3.34. Influence of antioxidants on the onset temperature of oxidation of a polyether polyol. DSC: heating rate 5 °C/min, oxygen atmosphere

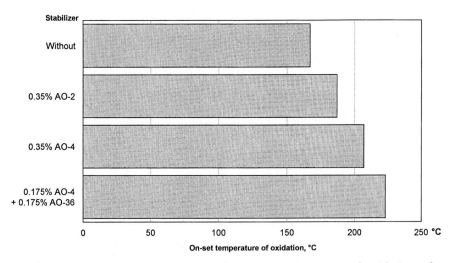

Fig. 3.35. Influence of antioxidants on the onset temperature of oxidation of a polyester polyol. DSC: heating rate 5 °C/min, oxygen atmosphere

tive conditions. The effect of an antioxidant in inhibition of thermooxidation of a polyol is determined by thermoanalytical methods such as DTA or DSC, while, e.g., at constant heat-up rate, the temperature at which exotherm oxidation occurs is measured. Figure 3.34 depicts the effect of some antioxidants on the on-set temperature of exothermy of polyether polyols, and Fig. 3.35 that on polyester polyols.

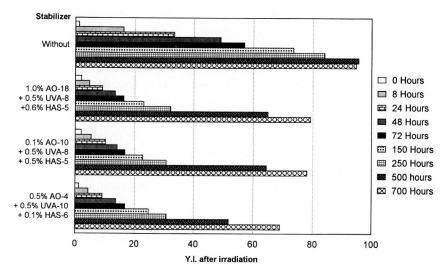

Fig. 3.36. Influence of stabilizers on the discoloration, Y.I., of thermoplastic polyurethane. Sample size: 50 μm cast films. Exposure device: Weather.O.Meter CI 65, b.p. temp. 63 °C, dry

Blends of the sterically hindered phenol, AO-4, with the aromatic amine, AO-36, yield the best results and lead generally to reduced discoloration under thermal stress. The use of the phenolic antioxidant AO-2 may lead to yellowing of the substrate or, in contact with other plastics, cause discoloration of the contact substrate by migration.

Stabilization of polyurethanes against photooxidation is primarily required for materials such as textile coatings or synthetic leather. Combinations of phenolic antioxidants with UV absorbers and sterically hindered amines, HAS, are particularly suitable [215]. Figure 3.36 presents a summary of the results obtained from light stabilizing experiments on T-PUR using such combinations [216].

To achieve good results, relatively high stabilizer concentrations have to be used. A prerequisite is that the stabilizers are compatible with the polymer matrix to prevent blooming.

3.10
High Performance Engineering Thermoplastics

The principles outlined regarding stabilization of, among others, polyolefins, polystyrene, polyamides, polyester, polyoxymethylene and polycarbonate are valid for all engineering thermoplastics. Sterically hindered phenols, and phosphites or phosphonites protect the polymer during processing and thermal long term aging. Application of light stabilizers such as UV absorbers or sterically hindered amines contributes to stabilization against light induced degradation. The choice and concentration of suitable stabilizers and stabilizer blends depends on their efficiency in the substrate to be tested. At the same time, it has to be ensured that no plateout occurs and that the stabilizers have no antagonistic effect. Because engineering plastics such as poly (phenylene ether) and polysulfones are processed at elevated temperatures, thermal stability of the stabilizers at these temperatures and their volatility is of the greatest importance. Whilst most thermally stable polymers have inherently good properties regarding thermooxidative stability, protection against light-induced degradation is necessary. So far, few results are available concerning stabilization of such polymers. Recent investigations concerned with the stabilization of poly (phenylene ether) [217] have shown that low-volatility UV absorbers, sterically hindered amines and combinations of sterically hindered amines with UV absorbers yield the best results with regard to discoloration under light impact.

3.11
Polymer Blends and Alloys

The same considerations apply to the stabilization of polymer blends and alloys. The most reactive phase regarding oxidative degradation is preferentially addressed and should, consequently, be protected by suitable stabilizers.

Because, to date, there is no knowledge available regarding distribution of common stabilizers in multiphase polymers, statements concerning efficiency of stabilizers and stabilizer blends are difficult to make. Only aging experiments can optimize their effect. To localize stabilizers in the phase to be stabilized, they are grafted onto that phase [218, 221].

Thus, Gilg and co-workers [222] succeeded in grafting sterically hindered amines onto the butadiene phase of blends based on polycarbonate and ABS, in this way providing focused protection against photooxidative degradation. Consequently, HAS bonded to the BR phase cannot exert antagonistic effect on the PC phase [223].

The focused stabilization of different phases in blends and alloys can be expected to gain importance in future developments.

CHAPTER 4

Influence of Metals, Fillers and Pigments on Stability

4.1
Metal Ion Deactivators

Metal ions catalyze the decomposition of peroxides forming reactive radicals and thus contribute to accelerated autoxidation (Eqs. 1.16 and 1.17). Deactivation of this process by suitable metal ion deactivators was proposed long ago by Downing and co-workers [224]. Hansen and co-workers [225] have shown the significant effect of copper on degradation of polypropylene and Hawkins and co-workers [226] arrived at similar results for polyethylene in contact with copper. Allara and Chan [227] were able to demonstrate that the oxidation state of the metal ion plays an important role in the degradation of polyethylene. Allara and co-workers [228, 229] have also shown that the catalytic reaction occurs at the interface between copper and the polyolefin. Later on, Sack and co-workers [230] and Wagner and co-workers [231] found that the catalytic effect of copper ions also extends further into the polymer. Therefore, it can be assumed that in addition to heterogeneous reactions at the polymer/metal interface, there are also homogeneous reactions of migrating copper ions inside the polymer itself.

The development of stabilizers capable of suppressing the reaction between metal ion and peroxides was a prerequisite for the use of polyolefins in contact with metals. This applies particularly to copper in the manufacture of cables for the electrical industry. Most investigations were carried out in the laboratories of the cable industry. A summary of the essential work was published by Chan [232].

The efficiency of such stabilizers is based on their ability to form stable complexes with metals and in particular, with copper ions [227, 233, 234], as shown in Fig. 4.1. In the literature they are generally referred to as metal deactivators [235].

A large number of metal deactivators are mentioned in the literature. However, there are only few commercially available stabilizers used in practice. These are applied for the stabilization of solid polyolefin insulations, cellular polyolefin insulations, crosslinked polyethylene insulations as well as for the stabilization of rubbers.

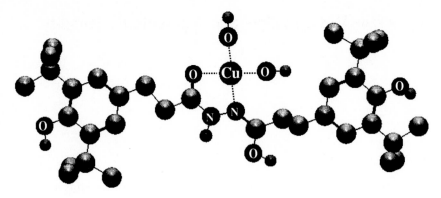

Fig. 4.1. Cu^{2+}-complex with metal deactivator MD-1

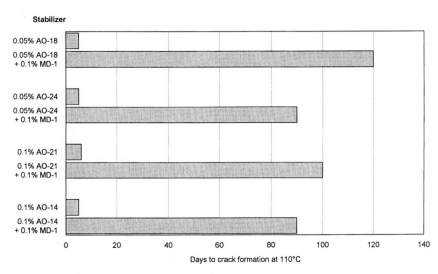

Fig. 4.2. Influence of a metal deactivator , MD-1, on the long term thermal stability (LTTS) of PE-LD wire insulation [238]. Copper conductor: 0.5 mm, insulation: 0.2 mm. Exposure device: circulating-air oven, 110 °C, days to cracking, according to VDE 0209, FTZ 72 TV1, draft

If hydrocarbon fillers, such as petrolatum are used as water-blocking compounds to render a cable waterproof, they must not be able to extract the stabilizers [236]. In the production of foamed insulations, metal deactivators must not react with the blowing agent thereby losing their effectiveness [237].

Figure 4.2 represents a summary of the results obtained from experiments concerned with thermal long term stability of telecommuni-

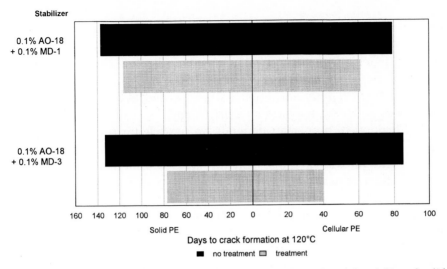

Fig. 4.3. Influence of metal deactivators on the long term thermal stability of solid and cellular PE-HD wire insulation [235]. Copper conductor: 0.65 mm, insulation: 0.35 mm. All samples contain: 1% TiO$_2$. Petrolatum treatment: 1s dipped in petrolatum at 115 °C, preaged at 70 °C for 10 days. Exposure device: circulating-air oven, 120 °C, days to cracking, according to VDE 0209, FTZ 72 TV1, draft

cation cables with a copper conductor having a diameter of 0.5 mm and insulation layer of PE-LD of 0.2 mm thickness.

The significant contribution of the metal deactivator to long term thermal stability of such a cable is clearly demonstrated. Communication cables are often filled with petrolatum to improve waterproofness. Figure 4.3 shows the effect of treatment with petrolatum on long term thermal stability of PE-HD cables having solid and cellular insulation. It is clearly demonstrated that treatment with petrolatum reduces the long term thermal stability because of extraction of the stabilizers from the insulation.

The cellular insulation has lower thermal stability than the solid insulation. So far it has not been established whether this is due to the influence of the blowing agent [237] or to higher oxygen diffusion [235] in the foamed structure.

The two examples demonstrate the importance of using metal deactivators in applications of plastics in contact with metals. This applies also to plastics/metal combinations such as fittings. Moreover, the addition of metal deactivators when mineral fillers with metallic impurities are used may lead to improvement of long term thermal stability, particularly of polyolefins.

4.2
Fillers

Fillers, particularly based on talcum and chalk, but also wollastonite and mica can alter the properties of the polymer in many ways. Thus, they raise, e.g., rigidity and impact strength, and improve dimensional stability by reducing warpage and shrinkage.

Modification of properties depends on particle form and size of the fillers. Their dispersion ability is controlled by specific surface, surface energy, and functional groups on the surface as well as by the way they are incorporated into the polymer. The presence of fillers in the polymer influences its long term thermal stability and its behaviour under photooxidative conditions. Klingert [239] and Fay [240] investigated the behaviour of talcum, and chalk-modified polypropylene under thermal and photooxidative conditions. They have shown that the nature of the filler has great influence on aging stability. They also found that fillers can adsorb stabilizers and that suitable treatment of the surface with a filler deactivator, FD, increases the effect of the stabilizers, particularly regarding long thermal stability. Figure 4.4 shows the influence of an epoxy resin based filler deactivator (Filler deactivator:Epoxy resin Araldite 7072, Araldite is a registered trademark of

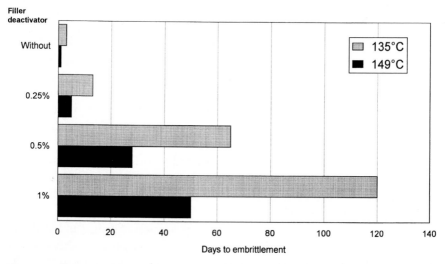

Fig. 4.4. Effect of a filler deactivator on the long-term thermal stability (LTTS) of 40% talc filled polypropylene. All samples contain: 0.1% AO-18, 0.3% TS-1 and 0.1% Ca-stearate. Sample size: 1 mm compression molded plaques. Exposure device: circulating-air oven

Ciba Ltd.) on long term thermal stability of polypropylene containing 40% talcum [238].

Deactivation of the filler's surface with epoxy resin successfully prevents adsorption of stabilizers. To achieve best possible protection against light-induced degradation, sterically hindered amines, HAS, are preferably used. Combinations of low and high molecular weight HAS have shown the best effect [239].

Naturally, the properties of the filler, especially metal ion contamination, have great influence on the long term behaviour of the filled material. Fillers with special surface coatings are commercially available and generally display better performance than comparable untreated materials. The addition of carbon black for coloration of filled materials can also lead to deterioration of long term stability by adsorption of the stabilizers onto the surface of the carbon black particles. Here too, deactivation of the particles' surface brings about clear improvements [239]. Developments concerned with the manufacture of fillers with specific deactivation of the surface, such as by coating or suitable functional groups may lead to further improvements of filled materials.

4.3
Pigments

Pigments are used to color plastics. Pigments are by definition colored particles insoluble under processing conditions, with a primary particle size of ~0.01–1 µm. Both organic and inorganic pigments are being used. Special treatment of the pigment's surface, e.g. by coating, leads to improvement in dispersion ability (wetting of the surface) in a given plastics material. Within the context of this monograph, of greatest interest is the influence of pigments on the thermal and photooxidative behaviour of the polymer.

4.3.1
Non-Colored Pigments

Originally, carbon black was used as light stabilizer. The effect depends on carbon black concentration as well as on the thickness of the substrate. A concentration of 2.5% is mentioned in the literature as optimal [241].

Fay and King [242] have shown that the light stability of polypropylene containing different amounts of carbon black can be significantly

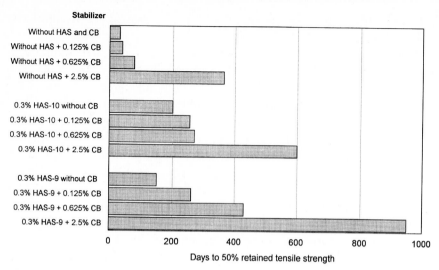

Fig. 4.5. Influence of HAS on the light-stability of a PP-homopolymer containing different amounts of a carbon black (CB). All samples contain 0.1% AO-18, 0.1% PS-2 and 0.1% Ca-stearate. Sample size: 2 mm injection molded plaques. Exposure device: Weather.O.Meter, b.p. temp. 63 °C. (CB = Vulcan 9, Vulcan is a registered trademark of Cabot Ltd.)

improved by the addition of sterically hindered amines, HAS, as shown in Fig. 4.5.

The influence of carbon black on long term thermal stability is depicted in Fig. 4.6.

The use of carbon black considerably reduces long term thermal stability. Gilg [243] has shown that the characteristics of carbon black, e.g. primary particle size, specific surface area, structure and surface chemistry, have a decisive influence on the behaviour of the plastics material. Light stability as well as long term thermal stability can be improved by the addition of sterically hindered amines, HAS. For long term thermal stabilization of carbon black-containing polyolefins, combinations of phenolic antioxidants such as AO-3 together with a thiosynergist such as TS-1 can also be used [243].

Rotschova and Pospisil [244] have found that the addition of metal deactivators such as MD-1 can significantly improve the effectiveness of carbon black containing metallic impurities with regard to thermo-oxidative stability. For white pigmentation, TiO_2 is the preferred material. Because titanium dioxide in its anatase modification accelerates the photooxidative degradation of the polymer matrix, particularly of polyolefins due to its high oxidation potential, rutile is usually used. Titanium dioxide particles that have been treated with, e.g., a coating

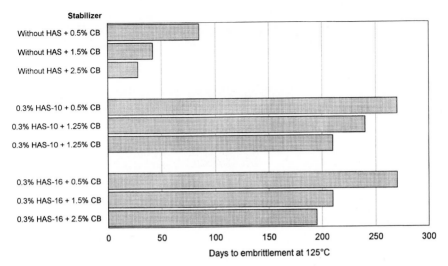

Fig. 4.6. Influence of HAS on the long-term thermal stability (LTTS) of a PP-homopolymer containing different amounts of a carbon black (CB). All samples contain 0.1% AO-18, 0.1% PS-2 and 0.1% Ca-stearate. Sample size: 2 mm injection molded plaques. Exposure device: circulating-air oven. Carbon black-type: Vulcan 9

of silicone along with inorganic Zn/Al modifiers, substantially increase the weathering resistance of TiO_2 – pigmented PE-LD films [245]. The combination of titanium dioxide with sterically hindered amines, HAS, further contributes to the light stability of white-pigmented articles [242].

4.3.2
Color Pigments

Mass coloring of plastics articles with color pigments is practised world-wide. Organic and inorganic color pigments are being used. A summary of the application of such pigments has been published by Herrmann and Damm [246].

Pigments such as different ultramarines and phthalocyanines are widely used. Many of the pigments contain, however, heavy metals such as cadmium, chromium or nickel. The use of these pigments is being questioned for environment-protection reasons.

Therefore, today organic pigments without heavy metals are preferentially used. Attention has to be paid to the thermal stability of organic pigments because of the high processing temperatures. Furthermore, pigments can act as nuclei in partially crystalline polymers such as polyolefins. This can lead to warping in the production of injection

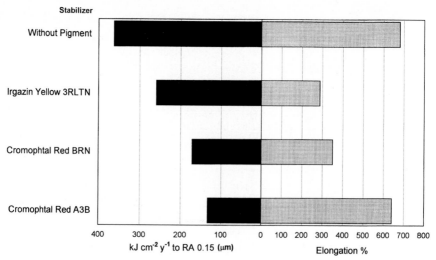

Fig. 4.7. Influence of some pigments on the light stability, mean surface roughness (RA) of PP-plaques after exposure to light and on the elongation after injection moulding. Pigment concentration: 0.1%. All sample contain 0.05% AO-18, 0.1% PS-2 and 0.05% Ca-stearate. Sample size: 2 mm injection molded plaques. Exposure: Florida, 45 °C south. (Color Index No.: Irgazin Yellow 3 RLTN: Pigment Yellow 110, Chromophtal Red BRN: Pigment, Red 144, Chromophtal Red A3B: Pigment Red 177)

moulded parts. The suitability of various pigments having different chemical structure regarding their use for mass coloring is generally described in the technical literature published by the manufacturers of pigments and should be strictly observed. Pigment preparations are being used to improve dispersion ability in the plastics melt. Within the framework of this monograph, of particular interest is the influence of organic pigments on the thermal, and photooxidative stability of plastics.

Figure 4.7 depicts the influence of some organic pigments on the photooxidative behaviour of polypropylene [247]. Polypropylene plaques differ with regard to elongation after injection moulding, thus leading to the conclusion that this is due to the pigment's nuclei-forming influence.

The results show that pigments have differing influence on the photooxidative behaviour of PP plaques.

The use of stabilizers significantly improves the stability of the plastics material against light degradation. Figure 4.8 shows the influence of different light stabilizers on aging of red pigmented PP plaques [247].

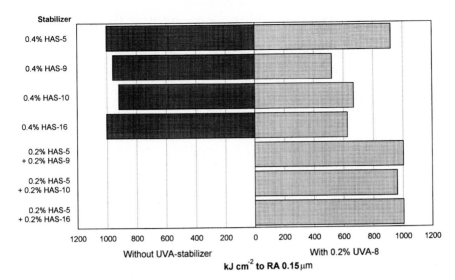

Fig. 4.8. Influence of stabilizers on the light stability, mean surface roughness (RA), of red pigmented PP-plaques. Pigment concentration: 0.1%, Cromophtal red BRN. All sample contain: 0.05% AO-18, 0.1% PS-2 and 0.05% Ca-stearate. Sample size: 2 mm injection molded plaques. Exposure: Florida, 45 °C south

The results clearly indicate that sterically hindered amines, HAS, contribute better by themselves to protection against photooxidation than UV absorbers alone.

The combination of sterically hindered amines, HAS, with a UV absorber makes almost the same contribution as HAS alone, although in this case at half the concentration of HAS. Optimal protection is achieved by a combination of a UV absorber with a blend of a low molecular weight and a high molecular weight sterically hindered amine, HAS. Such combinations also contribute to long term thermal stability of the pigmented substrates.

Hinsken and Meyer [248] have found that certain oligomeric sterically hindered amines like HAS-10 can decrease the colour strength of pigmented articles. This effect can manifest itself in the manufacture of mass colored fine PP fibres, where relatively high pigmentation is necessary to achieve the desired colour intensity and brilliance [249]. Zweifel [250] has found that nuclei formation by pigments can be partially suppressed by interaction with sterically hindered amines such as HAS-10, HAS-11 and HAS-14. Horsey and co-workers explain this effect by agglomeration of the pigments under special processing conditions [251]. This interaction with pigments was not observed in using the sterically hindered tertiary amine HAS-16 [249, 251]. The relationships shown here concerning improvement of the light stability of pigmented plastics parts by the addition of suitable sterically hindered amines, HAS, and UV absorbers are valid for practically all organic pigments. However, attention should be paid that the pigments do not contain metallic impurities. The latter act as catalytic hydroperoxide decomposers and thus negatively influence stability under photooxidative conditions.

CHAPTER 5

Solubility and Diffusion of Stabilizers in Polymers

To inhibit autoxidation efficiently, a stabilizer has to be dissolved in the polymer as homogeneously as possible and in sufficient concentration commensurate with its effectiveness – in general, at most 1–2 wt%. Because commonly used stabilizers contain polar groups, and are thus dipolar, solubility is strongly influenced by the nature of the polymer matrix and of the stabilizer itself. Polymers with polar functional groups such as polyesters, polyamides, poly (methyl methacrylate), polycarbonate or polyacetal provide a suitable environment for the stabilizer's solubility. One has to take into account that addition of relative high concentration of a low molecular additive may lead to a decrease in the T_g of the polymer.

Stabilizers are present only in the amorphous part of the polymer [252] because during crystallization of the polymer they are rejected into the amorphous area at the phase interface of the growing crystal. The local concentration in the amorphous region can be, therefore, several times higher than the original stabilizer concentration. Because, as shown, oxygen is dissolved only in the amorphous area, the stabilizer is present just where it is needed to suppress oxidative degradation.

The situation is quite different with the non-polar and partially crystalline polyolefins. Here, the polar stabilizers do not have an environment ensuring compatibility by a variety of interaction effects. Again, the stabilizer molecule is dissolved only in the amorphous area [253]. If the stabilizer concentration is higher than its solubility, then its molecules can leave the polymer by diffusion to the surface or phase separation between stabilizer and polymer may occur. Knowledge of the solubility and diffusion behaviour of a stabilizer is a prerequisite for its correct selection.

5.1
Solubility in Theory

The solubility of a relatively small molecule in a given polymer is the concentration that is in equilibrium with the pure solute at the same temperature and pressure. The theoretical treatment of the solubility of small molecules, based on the regular solution approach introduced by Gee [254], has been discussed in depth by Roe and co-worker [255] and by Billingham and co-workers [256–258]. For a crystalline stabilizer in contact with the polymer's surface, solubility is defined by the condition that the negative free energy of mixing of the liquid stabilizer with the polymer is equal to the positive free energy required to convert the crystalline stabilizer to liquid at the same temperature. Combined with the Flory-Huggins theory of the free energy of mixing [259] and with the assumption that the solubility of the stabilizer is low, the following equation results:

$$-\ln S = \frac{\Delta H^f}{RT}\left[1 - \frac{T}{T_m}\right] + \left[1 - \frac{V_1}{V_2}\right] + \chi \qquad (5.1)$$

S is the solubility of a stabilizer with melting point, T_m, heat of fusion, ΔH^f, and the molar volume V_1. V_2 is the molar volume of the polymer in question and χ represents a polymer/stabilizer interaction parameter.

According to the regular solution model [258], the solubility of a stabilizer in a given polymer is governed by three factors.

(i)   The first results from the free energy of fusion of the stabilizer at a chosen temperature, T, and is positive for $T<T_m$, if $T>T_m$, this term becomes zero. Crystals with high ΔH^f are expected to be less soluble than crystalline modifications with lower melting point. That solubility is affected by the crystal morphology has been shown by Földes and Turcsanyi [260].

(ii) The second term arises from the geometric entropy of mixing of stabilizer and polymer.

(iii) The third term represents a measure of the compatibility of the stabilizer-polymer combination but in a strict sense a stabilizer is only compatible with a given polymer if χ is negative.

Billingham and co-workers [257, 258] have shown that the application of Eq. (5.1) to the solubility of stabilizers in polypropylene is qualitatively valid. However, it cannot be used to predict the solubility of a given stabilizer since χ values are unpredictable. Modelling and computing solubility is a difficult problem, because the theory is strictly applicable only to the liquid state [261].

While the regular solution model approach concentrates on the role of stabilizer structures, Shlyapnikova and co-workers [262–264] developed the concept of the "site adsorption model" to explain the solubility of low molecular weight compounds in polypropylene. This model focuses on the role of the polymer in determining solubility. On the basis of this model, the bulk of the solute is reversibly adsorbed at the topological sites of increased free volume present in any polymer, and only small amounts of the solute are molecularly dispersed in the 'normal' polymer. However, interpretation of the form of temperature dependence is similar to that predicted by the regular solution model [265].

5.2
Solubility in Practice

Different methods are described in the literature for the measurement of solubility of low molecular weight compounds in polymers. Solubility can be determined by measuring the equilibrium uptake of stabilizer into the film of a given polymer if the experiments are sufficiently long to ensure full equilibrium [257, 258]. A different approach has been chosen by Moisan [266, 267]. In this case, the solubility is determined by extrapolation of diffusion data to infinite time.

In general, data obtained by the two different methods are in a good agreement.

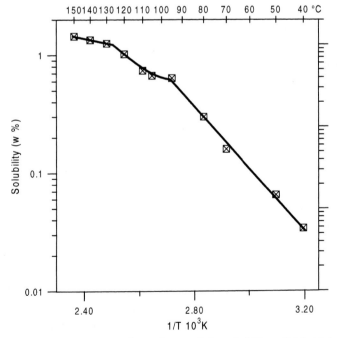

Fig. 5.1. Temperature dependence of the solubility of AO-18 in polypropylene

Table 5.1. Solubility properties of some stabilizers

Stabilizer	M_w °C	T_m °C	Polymer	Temp. °C	S w/w %	Ref.
AO-2	222	69	PP	70	2.27	268
				25	0.44	268
AO-18	1178	125*	PP	70	0.40	268
				25	0.074	268
				25	0.014	272
UVA-1	228	62	PP	70	2.10	268
				25	0.27	268
UVA-2	326	47	PP	70	4.36	268
				25	0.67	268
UVA-6	225	132	PP	30	0.07	269
AO-3	531	55	PE-LD	30	0.58	266
AO-14	545	190	PE-LD	45	0.26	267
				30	0.014	267
AO-18	1178	125*	PE-LD	80	0.011	260
	1178	125*	PE-LD	30	0.007	266
			PE-LLD	25	0.012	272
HAS-5	481	86	PE-LD	30	0.125	270
HAS-10	Oligomer		PE-LD	30	0.01	271

* T_m dependent on crystal modification

Table 5.1 shows the solubilities of different stabilizers in various polymers, including data extrapolated to ambient temperature.

For AO-18 , the values reported differ to a large extent because of the fact that this stabilizer crystallizes in different crystal modifications [260]. Figure 5.1 shows the temperature dependence of the solubility of AO-18 in polypropylene [272]. The change of the slope above 126 °C and below ~95 °C is caused by the melting of the stabilizer.

Since antioxidant AO-18 can be present in various crystal modifications (a – δ) with defined melting points [260], the temperature range in which the slope changes is rather large.

Conclusions

The equilibrium solubilities of stabilizers in polyolefins at ambient temperatures are extremely low. Because their heats of solution are generally high, solubility increases rapidly with temperature. Therefore, in the concentrations normally used, the stabilizers will be totally soluble in the molten polymer at processing temperatures. Two cases may arise by cooling down the melt to ambient temperatures.

1) The stabilizer concentration is below equilibrium solubility at ambient temperatures. The total amount of stabilizer present remains soluble in the polymer at ambient temperature and, therefore, forms a molecularly dispersed solution. The only driving force for the stabilizer to move to the polymer's surface is by evaporation from the surface layer or being dissolved by a solvent in contact with the polymer's surface ('leaching').

Fig. 5.2. Surface precipitate of a stabilizer caused by metastable, supersaturated state at ambient temperature

2) The stabilizer concentration is above equilibrium solubility at ambient temperature. If the stabilizer concentration is above its equilibrium solubility at ambient temperature, it should precipitate on cooling down the polymer melt. The high viscosity of the polymer suppresses crystallization of the stabilizer in the bulk and the stabilizer remains in a molecularly dispersed, metastable, supersaturated state. This state should be followed by a precipitation within the polymer matrix. However, practice shows [268] that the stabilizer may find it easier to migrate to the surface and produce a precipitate there ('blooming'), as shown in Fig. 5.2.

If the stabilizer remains in the polymer at a concentration below saturation, the only means of loss from the surface is evaporation or, when in contact with a solvent, by leaching. If the stabilizer is in a metastable, supersaturated state, then the stabilizer is also lost from the surface layer by blooming.

5.3
Physical Loss of Stabilizers from Polymers
Through Diffusion and Evaporation

As a result of physical loss of the stabilizer, autoxidation of the polymer can no longer be inhibited. Further, plastic parts protected by suitable stabilizers for long term useful lifetimes may be exposed to permanent contact with substances that are better solvents for the stabilizers than the polymers into which they are incorporated.

Many polymers are used as packaging for food. Polymer parts, e.g. toys, household items, contain stabilizers needed during processing and as long term protection.

For this reason, it has to be ensured that stabilizers used are toxicologically harmless and that they do not enter foodstuffs by diffusing in too high concentrations out of the polymer. Regulations have, therefore, been issued concerning the use of stabilizers in indirect contact with food[1], e.g. food packaging materials. It follows that knowledge of diffusion processes of stabilizers in polymers is mandatory. The diffusion of gases, e.g. their permeation of polymers has been, and is, the subject of many theoretical and practical studies [255, 261, 266, 267, 273].

---

[1] Directive 90/128/EEC and its amendments 92/39/EEC, 93/9/EEC and 95/3/EEC. Code of federal regulations 21 CFR 170.3, parts 170 to 199 (April 1. 1995)

5.3.1
Stabilizer Loss in Theory

The physical loss of stabilizers (or, generally, of low molecular compounds) from a polymer matrix has been fundamentally described by Calvert and Billingham [274].

In general, the stabilizer is lost by dissolving into a flowing medium in contact with the polymer surface. The flowing medium could be a gas, e.g. air, or a liquid. The concentration of the stabilizer directly on the surface is, therefore, nil and has to be replenished continuously with new stabilizer molecules by diffusion from the polymer matrix. A mathematical model describing this situation needs two parameters: a mass transfer parameter characterizing transfer across the surface and a constant characterizing mass transfer within the bulk of the polymer, related to the diffusion coefficient, D. Additive loss is similar to radiation heat loss with the two parameters: surface emissivity and thermal conductivity respectively.

5.3.1.1
Loss of Stabilizer by Evaporation

If a stabilizer is molecularly dispersed in a given polymer at a concentration below its saturation solubility at ambient temperature and in contact with air, loss can only occur by evaporation from the polymer surface [275]. With the assumption that the volatility of the stabilizer from the polymer surface is proportional to its vapour pressure, and that Raoult's law is valid at all concentrations, the rate of mass transfer across unit area of surface is

$$\frac{dm}{dt} = \frac{V_o \cdot C_s}{S} = HC_S \tag{5.2}$$

where V_o is the volatility of the pure stabilizer under given test conditions, C_s is the concentration of the stabilizer at the surface, and S is the saturation solubility of the stabilizer under the same conditions. The parameter H is the required mass transfer constant and can be calculated by measuring V_o and S.

Loss from the surface creates a concentration gradient and the stabilizer lost by evaporation will be replaced by its diffusion from the bulk. Mass transfer within the polymer is described by Fick's laws of diffusion:

$$\frac{dm}{dt} = -D\frac{dC}{dx} \tag{5.3}$$

Table 5.2. Limiting equations for loss-times of stabilizers

Sample Size	L - Value (L = $H l/D$)		Time to 90 % Stabilizer Loss	Mode of Loss
Film (Thickness $2l$)	High L	(L>10)	t= 0.87 l^2/D	Diffusion
	Low L	(L<0.6)	t= 2.42 l/H	Volatility
Fiber (Radius l)	High L		t= 0.35 l^2/D	Diffusion
	Low L	(L<0.3)	t= 1.12 l/H	Volatility
Bulk	High L	(L>3)	t= 32 l^2/D	Diffusion[*]
	Low L		t= 30 D/H^2	Volatility

[*] Also valid for blooming to saturation equilibrium!

where D is the diffusion coefficient of the stabilizer and dC/dx is its concentration gradient. Equations (5.2) and (5.3) can be combined and solved with appropriate boundary conditions. The solution for different cases and sample geometries has been described by Crank [276]. Transport equations describing the loss of a low molecular weight compound where both evaporation and diffusion within the bulk may be rate controlling have been summarized by Calvert and Billingham [274].

Their solutions are presented as equations relating to the time for 90% of a stabilizer to be lost to the dimensionless parameter L=Hl/D where l is the radius of a fibre, the half thickness of a film or the depth at which concentration falls to 10% of its initial value in a semi-infinite solid. The loss of stabilizers from thick samples (slow diffusion, rapid evaporation) is diffusion controlled and Hl/D is a large value. If, however, loss of stabilizer is controlled by evaporation (rapid diffusion and slow evaporation) Hl/D becomes small [275]. Estimates of additive properties suggest that in this case evaporation controlled loss for thin films and fibres takes place and diffusion controlled loss in the case for thicker samples. The limiting equations of Table 5.2 also provide a method for testing the rate controlling mechanism by measuring the dependence of loss rate on sample dimension.

5.3.1.2
Loss of Stabilizer by Surface Precipitation

If the polymer is surrounded by a medium which is a good solvent for the stabilizer instead of by air, then the rate of removal of the stabilizer from the polymer's surface is very high. The rate-determining step is the stabilizer's diffusion in the polymer, independent of the geometry of the part. A similar situation arises if the stabilizer is in a metastable, supersaturated state in the polymer. If precipitation of the stabilizer occurs on the polymer's surface, and the precipitate remains there, then the final concentration of the remaining stabilizer reaches saturation solubility at a rate determined by diffusion.

The loss process is driven only by diffusion through a concentration gradient in the sample. The only parameter required is the diffusion coefficient, D. The stabilizer concentration can only fall below its saturation concentration if the precipitate is removed from the polymer surface, e.g. mechanically.

A more complex situation arises if the solvent can penetrate the polymer. The rates of diffusion of the stabilizer in the swollen and unswollen polymer are different and may facilitate diffusion of the additive outwards. Such conditions may be important regarding migration of additives into oily food [277].

5.3.2
Diffusion in Practice

Numerous studies have been reported in the literature [255, 257, 258, 265, 269, 274, 278, 279] concerning the measurement of diffusion coefficients of various stabilizers. Luston's work [279] includes many volatility measurements with different stabilizers. A summary of a variety of data has been published by Flynn [280]. Most data were obtained with polyolefins as substrate. Values published by various authors vary substantially [17]. One reason is certainly the dependence of diffusion on the morphology of the polymer which has strong influence, particularly with partially crystalline polyolefins. All studies described here are based on diffusion behaviour following Fick's laws. It is assumed that the diffusion of small molecules like stabilizers proceeds in the amorphous area of the partially crystalline polymer. Anomalous diffusion behaviour can sometimes be explained by morphological characteristics [261, 273]. Generally, it is assumed that the diffusion coefficient is independent of concentration because of the small quantities used. Diffusion experiments should be carried out only in well-defined samples with known morphology. Table 5.3 shows a summary of some diffusion constants.

The solubility of a stabilizer is influenced insignificantly by the morphology of the polymer: it is inversely proportional to the polymer's crystallinity. Diffusion constants are influenced in a complex way by the crystallinity as well as by the morphology of the polymer.

Table 5.3. Diffusion coefficients of some stabilizers

Stabilizer	M_w °C	T_m °C	Polymer	Temp. °C	D cm²/s		Ref.
AO-2	222	55	PP	25	5	10^{-10}	268
AO-18	1178	125	PP	25	6.5	10^{-15}	272
UVA-6	225	132	PP	30	3.29	10^{-11}	269
AO-2	222	70	PE-LD	25	9.6	10^{-10}	281
AO-3	531	55	PE-LD	30	3.04	10^{-10}	266
				30	1.43	10^{-10}	282
AO-18	1178	125	PE-LLD	25	5.6	10^{-11}	272
			PE-LD	80	13.4	10^{-9}	260
			PE-LD	30	1.05	10^{-11}	266
HAS-5	481	86	PE-LD	30	9.42	10^{-10}	270
HAS-10	Oligomer		PE-LD	30	1.12	10^{-14}	271

In particular, diffusion requires co-operative motions of polymer and diffusant so that D is expected to decrease with increasing density of the polymer. The diffusion of small molecules in polymers can be treated on the basis of the fractional free-volume concept. Billingham has reviewed the different approaches to explain effects of diffusant structure [283]. Földes [284] has found good correlation between the diffusion rate of AO-3 and the free-volume of the non crystalline phase in PE-LD. The rather rigid molecule of AO-14 was found to be less dependent on the free-volume changes of the polymer compared with AO-3 which has a flexible, long linear chain with similar molecular mass [285]. Therefore, diffusion of the flexible molecule will occur faster through the polymer matrix.

The dependence of the diffusion constant of AO-18 in polypropylene on temperature is depicted in Fig. 5.3

The slightly curved Arrhenius plot appears to be a universal phenomenon but would be expected for a process involving cooperative motions of additive and polymer at temperatures not too far from T_g.

Table 5.4 summarizes the diffusion coefficients of a homologous series of sterically hindered phenols of the AO-3 type and of AO-18 as a compound with higher molecular weight in PE-LD as substrate.

Figure 5.4 shows the dependence of the difusion, D, on the molecular weight of a series of phenolic antioxidants as listed in Table 5.4.

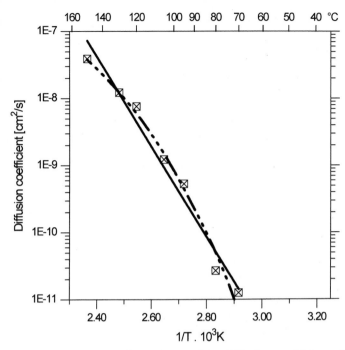

Fig. 5.3. Temperature dependence of the diffusion coefficient of AO-18 in polypropylene

Table 5.4. Dependence of diffusion coefficients upon molecular weight of hindered phenols in PE-LD at 30 °C

Structure	n	Mw g/mol	Diffusion Coefficient D_{30}	cm²/s	Ref.
	0	292	5.72	10^{-10}	282
HO—⟨ring⟩—(CH₂)₂–C–O–(CH₂)ₙ–CH₃	2	320	4.78	10^{-10}	282
	5	362	4.09	10^{-10}	282
	11	446	2.97	10^{-10}	282
AO-3	17	530	1.43	10^{-10}	282
AO-18		1178	1.05	10^{-11}	272

In Table 5.5 are listed the diffusion coefficients of several 2-(2-hydroxyphenyl) benzotriazoles having differing structures and molecular weights in PP as substrate.

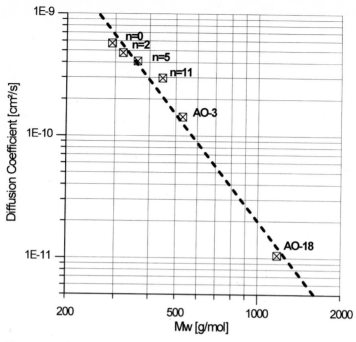

Fig. 5.4. Molecular weight dependence of diffusion coefficient for hindered phenols in PE-LD at 30 °C

Table 5.5. Dependence of diffusion coefficients upon molecular weight of benzotriazole UV absorbers in PP at 80 °C

Structure	Mw g/mol	Fp	Diffusion Coefficient D_{80} cm²/s	Ref.
UVA-6	225	132	$2.43 \cdot 10^{-8}$	269
UVA-7	323	106	$3.19 \cdot 10^{-9}$	269
UVA-11	351	83	$6.21 \cdot 10^{-9}$	269
UVA-14	448	141	$1.33 \cdot 10^{-9}$	269
UVA-15	659	198	$7.71 \cdot 10^{-10}$	269
UVA-16	761	114	$1.06 \cdot 10^{-9}$	269

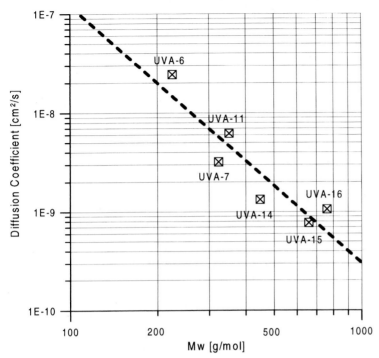

Fig. 5.5. Molecular weight dependence of diffusion coefficient for 2-(2-hydroxy-phenyl)benzotriazoles in PP at 80 °C

Figure 5.5 shows the dependence of the diffusion, D, on the molecular weight of this series of UV absorbers as listed in Table 5.5.

Within both series, the diffusion coefficient correlates roughly with the square of the molecular weight. The molecular weight dependence of the diffusion coefficient within the series of phenolic compounds has been analyzed in detail by Möller and Gevert [282]. Linear *Arrhenius* plots were obtained. The activation energy E_a increased linearly with increasing molecular size. Billingham has reviewed in detail the different approaches to explain the effects of diffusant structure [286].

Conclusions

The diffusion of stabilizers in polymers follows Fick's laws. At the low concentration used, diffusion coefficients are concentration-independent. In semicrystalline polymers, diffusion is strongly influenced by the polymer's total crystallinity and morphology. For a given stabilizer, D is in general lower in polypropylene and high density polyethylene compared with low density polyethylene. For a given molecular weight, molecules with flexible structures diffuse more rapidly than molecules with rigid and compact structures.

5.3.3
Stabilizer Loss in Practice

Solubility, volatility and diffusion of a stabilizer in a given polymer matrix are the rate-determining factors for the loss of a stabilizer from the polymer matrix. If these factors are known, quantitative prediction can be made concerning time-related progress. The necessary equations for the solution are summarized in Table 5.2. The value of the dimensionless parameter L=Hl/D indicates whether the loss process is controlled by diffusion or volatility. The values summarized in Table 5.6, show the time for 90% of a stabilizer to be lost from a film with thickness 2l, or the depth at which the stabilizer concentration falls to 10% of the original concentration 1 mm below the surface of the bulk polymer, including stabilizers such as hindered phenolic antioxidants, UV-absorbers and hindered amine stabilizers, HAS. Comparison of the low molecular weight antioxidant AO-2 and the high molecular weight AO-18 shows the influence of molecular weight on volatility as well as on the diffusion coefficient. However, increase in molecular weight also decreases solubility. The two benzophenone-type UV absorbers differ only in the length of the alkyl chain in the 4-position. Significant change in volatility and solubility is observed. Overall, the result is a large difference in the H/D ratio. The initially evaporation-controlled loss from thin samples becomes diffusion-controlled for the bulk of the polymer. Comparison of the low molecular weight HAS-5 with the oligomeric high molecular weight HAS-10 shows clearly the reduction of loss caused by the

Table 5.6. Calculated loss time for stabilizers from PE-LD

Stabilizer	Mwg/ mol	Temp °C	V_0 g/cm².s	S g/cm³	D cm²/s	H/D cm⁻¹	LOSS TIME (Hours, Day, Year)				Ref
							Film 10µm	Film 100µm	Film 1µm	Bulk	
AO-2	222	25	6 10⁻¹⁰	1.5 10⁻²	9 10⁻¹⁰	40	16 H (V)	7 D (V)	30 D (V,Df)	10 Y (Df)	286
AO-18	1178	25	5 10⁻¹³	2 10⁻⁴	5 10⁻¹²	5 10²	6 D (V)	0.5 Y (V,Df)	13 Y (Df)	2000 Y (Df)	286
UVA-1	228	25	5 10⁻⁷	5 10⁻⁴	7 10⁻⁹	1.5 10⁴	30 S (Df)	1 H (Df)	4 D (Df)	1.5 Y (Df)	286
UVA-2	326	25	5 10⁻¹²	1 10⁻²	1.5 10⁻⁹	0.33	28 D (V)	270 D (V)	7.5 Y (V)	6000 Y (Df)	286
HAS-5	481	20	2 10⁻⁷	2 10⁻³	4 10⁻¹⁰	2.5 10⁵	10 Min (Df)	18 H (Df)	70 D (Df)	25 Y (Df)	271
HAS-10	Oligomer	20	8 10⁻⁸	1 10⁻²	4 10⁻¹³	2 10⁷	7 D (Df)	2 Y (Df)	200 Y (Df)	2500 Y (Df)	271

(V) = Volatility controlled
(Df) = Diffusion controlled

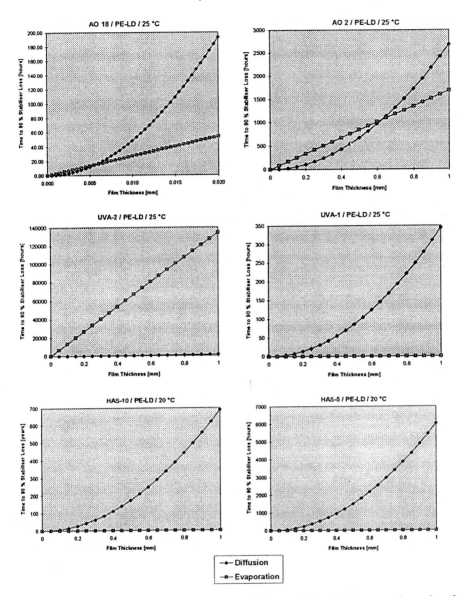

Fig. 5.6. Time to 90% stabilizer loss in relation to film thickness and mode of loss, evaporation and diffusion controlled, according to Table 5.6

stabilizer's molecular weight. The comparison of the loss mode – either by evaporation or by diffusion – is visualized in Fig. 5.6. Unfortunately, only few rates of evaporation involving the commonly used stabilizers have been measured so far.

Conclusions

The residence time of a stabilizer in a polymer decreases with decreasing thickness of the polymer. Even high molecular weight stabilizers have a relatively short residence time in very thin films. In fibres with small diameter, residence time would be even shorter (see Table 5.2). However, because fibres are usually highly stretched, the diffusion coefficient is much smaller than in the corresponding non-oriented polymers.

For all stabilizers it can be predicted that, in thick cross-sections, the rate limiting step is diffusion and that gradually, with decreasing thickness, volatility becomes rate-determining. These findings are valid for loss of stabilizer from polymers in contact with media in which the stabilizers are not, or only slightly, soluble, e.g. air. If the polymer's surface is in contact with a medium that is a better solvent for the stabilizer than the polymer matrix, then complete loss of the stabilizer can occur after a short time. For this reason, efforts are continuously made to develop stabilizers that cannot in practice diffuse out of the polymer matrix. As shown, this can be accomplished most easily by increasing the molecular weight, whereby compatibility and solubility should not be influenced too strongly. The manufacture and use of oligomeric stabilizers, e.g. HAS-10, or polymer-bound stabilizers represent a solution of the problem. Possible approaches in this respect are described in the literature [287, 288]. It should not be forgotten that stabilizers have to have certain mobility within the polymer if they are to react with the short-lived radicals formed by thermooxidation. Gugumus [289] has shown that oligomeric HAS derivatives reach an optimal effectiveness with increasing molecular weight, and lose it rapidly thereafter. Because of the interaction between stabilizer and substrate, the optimal molecular weight can differ [290]. Furthermore, the stabilizer has to be present primarily in sufficiently high concentration where it is supposed to be effective. For example, the polymer's surface is preferentially exposed to oxidative degradation. Consequently, a sufficiently high concentration of stabilizer is necessary in this region. Also, at least partially constant replenishment from the bulk of the polymer by migration is needed to compensate stabilizer consumed by the degradation processes. Blends of suitable stabilizers having different molecular weights can lead to synergistic effects.

Finally, it is decisive that the stabilizer remains in the polymer and because of its mechanisms protects the end-use article efficiently throughout its lifetime.

It should also be taken into account that morphology of the polymer and its polarity can change during aging caused by thermo- and photooxidative processes. Stabilizers too undergo changes as a result of their reactions with reactive oxidation products. Consequently, an unusually dynamic situation arises making predictions regarding lifetime very difficult.

5.4
Service Lifetime Predictions for Polymer End Use Applications

5.4.1
Theoretical Considerations

Based on the previously described relationships between the loss of a stabilizer and its diffusion and vapour pressure, conclusive statements regarding the lifetime of a stabilized polymer cannot be made. Malik and co-workers [291] have found that the physical loss of stabilizers that are not consumed over extended periods of time because they are regenerating themselves, e.g. sterically hindered amines, can be correlated with the lifetime of polymers subjected to light irradiation. Generally, however, complex systems are used for the stabilization of polymers. Most stabilizers are "consumed" in the described manner under the influence of oxygen, heat and light. The resulting lifetime of a plastics article depends, therefore, in a complex way on the kinetics of the polymer's degradation, diffusion of oxygen into the polymer matrix, and the effect of inhibition imparted by the stabilizer employed [292]. In the process, the stabilizer's molecules and the polarity of the polymer change during gradual oxidation and under heat, light and weathering influences.

Polymers having hydrolyzable groups on the backbone are also degraded by the effect of humidity. This has been shown by McMahon and co-workers with poly (ethylene terephthalate) [293] and several authors [294–296] with polycarbonate.

In these instances, oxidative degradation is superimposed on degradation by hydrolysis under weathering conditions.

Generally, degradation of the polymer starts under oxidative conditions when the stabilizer's molecules cease to suppress the autoxidation chain, because of purely physical loss, or by reaction with autoxidation products. In this connection it has been shown that certain transformation products of the originally used stabilizers are capable of inhibiting autoxidation. The effect of the stabilizers ceases when their concentration falls below a limit which, according to Shlyapnikov

[297] corresponds to a critical inhibitor concentration. In contrast, Gugumus [298] developed a model of the polymer's critical oxidation level. Concentrations valid for a given polymer and a given stabilizer have not been measured so far. It can be assumed that very low quantities are involved. Determination of this limiting value by analytical methods is practically impossible, because, as mentioned, certain transformation products of the original stabilizer act as inhibitors and, furthermore, blends of various stabilizers are used having complementary effects.

The objective of every stabilization of a plastic material is to protect it efficiently against oxidative degradation in a given medium for the desired lifetime. Examples of such media are, just to mention a few, outdoor weathering, e.g. bottle crates, window profiles, in the soil, e.g. cables, pipes, high temperatures, e.g. electronics, automotive applications, in contact with water, e.g. washing machines, sewage, or gasoline and oil, e.g. automotive, bottles. Useful lifetimes of up to 50 years or more can easily be demanded for various applications. Only a few studies can be found in the literature dealing with aging of polymers and their lifetimes in the course of long service. An investigation by Malek and Stevenson is based on a 100 year case study with natural rubber [299]. Unfortunately, in practically all publications only mechanical tests during aging [300] are mentioned without reference to stabilization and its influence.

The most informative studies regarding prediction of the service life of a plastic part are in the sector of cable insulations. Because power cables have to last for decades, their stabilization to achieve this objective is unavoidable. To make a prediction with regard to presumed lifetime, models have to be used that are based on extrapolation of laboratory results obtained under accelerated conditions such as elevated temperatures or increased irradiation intensity.

The preparation of test specimens, e.g. thickness, orientation caused by the preparation process or internal stress can also influence aging behaviour.

The proper choice of the test method to identify the failure criterion is of great importance. The degradation of plastics under oxidative conditions almost always proceeds in localized areas or centres. For this reason degradation is a heterogeneous process and its progress is better monitored by test methods measuring mechanical strength, elongation and toughness [301] than by monitoring material characteristics such as molecular weight and molecular weight distribution.

Moreover, it has to be taken into account that artificial aging of specimens subjected to heat and light yields results differing with re-

Table 5.7. Changes of material properties upon oxidative deterioration of polymers

PHYSICAL PROPERTIES	CHEMICAL PROPERTIES	MOLECULAR PROPERTIES
Mechanical Properties	**Spectral Properties**	**Microscopic Properties**
Tensile	• UV - VIS Spectroscopy	• Optical Microscopy
Flexural	• IR / Raman Spectroscopy	• UV - Microscopy
Impact	• Photoacoustic Infrared Spectroscopy	• Scanning Electron Microscopy
Hardness	• NMR Spectroscopy	• Transmission Electron Microscopy
Dynamic Mechanical Analysis	• ESR Spectroscopy	• Scanning Tunnel Microscopy
Surface Properties	• Mass Spectroscopy [TOF - MALDI]	• Atomic Force Microscopy
Roughness	• X-Ray Fluorescence	**Surface Properties**
Haze	• Atomic Absorption	• Auger Spectroscopy
Gloss	• Chemiluminescence	• Secondary Ion Mass Spectroscopy
Colour	**Chromatographic Properties**	**Free Volume Determination**
Thermal Properties	• Gas Chromatography	• Positron Annihilation Spectroscopy
Differential Scanning Calorimetry	• Liquid Chromatography	
Thermogravimetric Analysis	• Thin Layer Chromatography	
Thermomechanical Analysis	• Supercritical Fluid Chromatography	
Electrical Properties	• Size Exclusion Chromatorgaphy	
Dielectric Constant		
Mass Transfer Properties		
Migration		
Diffusion		
Evaporation		
Permeation / Solubility		

gard to long term performance from those obtained by outdoor weathering conditions where irradiation intensity varies with daytime and season. Thus during periods of darkness, UV stabilizers, for example, have the possibility to migrate from the interior of the sample and replace used-up stabilizers where they are most needed. If the polymer is subjected to stress during aging, e.g. tensile stress or compressive stress [302–304], the lifetime of the polymer can differ under these conditions from that of samples that have not been exposed to stress.

The most important parameters that can serve as indicators for the presumed lifetime of a plastic part under oxidative conditions are summarized in Table 5.7.

Recording of these data during aging and at different temperatures should permit comprehensive analysis regarding the expected service lifetime of a plastic material. The application of mathematical models such as the cumulative damage theory was described by Miner [305]. A summary of similar models for the prediction of lifetime was compiled by Shah and co-workers [306]. In view of the unrealizable experimental work involved in determination of all possible data describing changes of the material following oxidative aging (see Table 5.7), it is not surprising that simpler models are being used for the prediction of service lifetime. It suffices, e.g., that changes of mechanical properties are monitored in the course of aging. Determination of lifetime based on the simple Arrhenius relation and extrapolation of data obtained at relatively high temperatures or other drastic aging conditions is frequently used.

$$r_9 = r_0 e^{-\left(\frac{E_a}{RT}\right)} \tag{5.4}$$

In Eq. (5.4) r_9 is the rate constant of the chemical degradation reaction at temperature 9, r_0 is the rate constant of the reaction at a reference temperature (e.g. room temperature), E_a the Arrhenius activation energy, R the gas constant and T the absolute temperature. The basis of this equation is a linear relationship between the log of time and change of the measured material property at time t. Because the chemical reaction leading to changes of the material's property is usually not a straightforward reaction but rather a sequence of reactions, application of the Arrhenius relation is not permissible in a strict sense. [307, 308]. In particular, as the temperature is raised, reactions of high activation energy become particularly more important. Chan and co-workers [309] have shown that the Arrhenius relation still yields valuable insights for the prediction of service lifetime of copper cables stabilized with various metal deactivators. Gillen and Clough developed the principle of time – temperature superposition [310, 311] in view of the above-mentioned weaknesses of the Arrhenius relation and based on experimental findings from aging experiments with nitrile rubber. On condition that the Arrhenius activation energy E_a can be determined reliably, available data, in this case tensile elongation, can be correlated with a chosen reference temperature T_{ref}:

$$a_T = e^{\frac{E_a}{R}\left(\frac{1}{T_{ref}} - \frac{1}{T}\right)} \tag{5.5}$$

The condition for the application of this relationship is that the time-temperature superposition is satisfied. Should that not be the case, evaluation of the data in accordance with the Arrhenius relation is not possible and the test method or the failure criterion has to be redefined.

Crine has chosen a different approach [312–314] to aging of polymers under long term thermal conditions. It is based on the application of the rate theory associated with Eyring.

The time needed for the reactants to arrive at the reaction products through the energy barrier ΔG (*Gibbs* free energy) is

$$t \cong \frac{h}{kT} e^{\left(\frac{\Delta G}{kT}\right)} = \frac{h}{kT} e^{\left(\frac{-\Delta S}{k}\right)} e^{\left(\frac{\Delta H}{kT}\right)} \tag{5.6}$$

Equation (5.6), where h and k are *Planck's* and *Boltzmann's* constants respectively, can be used if a given material goes from its original state to another one, e.g. aged state, as it applies to thermal aging.

In this case, data obtained from aging are noted as log $(f\,T)$ vs $1/T$ and not as in the case of the Arrhenius relation as log t vs $1/T$.

5.4.2
Failure Criterion

Gillen and Clough [310] have shown with the example of thermal aging of nitrile rubber that time-temperature superposition (Eq. 5.5) using elongation as the failure criterion yields valuable results. However, if tensile strength is taken as failure criterion, the time-temperature superposition does not yield reasonable results. They were able to show that, in the course of thermal aging, the degradation in the interior of the sample proceeds faster than the diffusion of oxygen into the interior of the sample. Tensile strength is a material property that depends on the force at break integrated over the cross-section of the test specimen, and the aged samples, especially those aged at high temperature, showed little degradation in their interior regions. Oxidation on the surface of the samples was not influenced by diffusion-limited oxidation. If, instead of tensile strength, surface modulus is determined, agreement with time-temperature evaluation is possible.

Verdu [315] described the influence of aging of polymers on their mechanical properties. Aging of ductile materials leads to significant reduction of their original elongation and the sample becomes brittle.

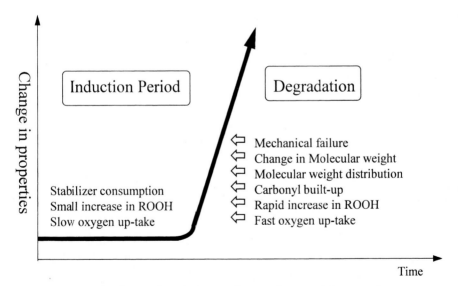

Fig. 5.7. Aging of polymers in relation to changes in material properties

In originally brittle materials, the difference after aging is far less pronounced. Here, aging leads in most instances to surface damage leading eventually to failure of the sample under mechanical stress.

It follows that choice of the most informative test method in determining the failure criterion is most important. Figure 5.7 shows schematically the relationship between the aging process and changes of the polymer's properties.

The duration of the induction period and the slope of the curve in the course of oxidative degradation depend on the chosen stabilization system and the concentration of the individual components. The influence of stabilizer concentration on lifetime has been investigated most thoroughly with polyolefins. Gugumus [316, 317] has shown that for polyolefins there is a linear relationship between stabilizer concentration [c] or [c$^{1/2}$] and lifetime. Audouin and co-workers [318] investigated the relationship between stabilizer consumption and service lifetime prediction of thermally aged crosslinked polyethylene and have shown that consumption of antioxidant is the first indicator for the subsequent degradation. However, analysis of the stabilizer system during aging is so complex that it cannot be carried out in practice. Determination of the hydroperoxide concentration has a lower analytical limit of 10^{-5} mol OOH/kg polymer. The use of chemiluminescense to identify preceding processes, primarily peroxide formation during the induction period, has recently shown promising results [319]. Further development will need, however, more basic research.

CHAPTER 6

Testing of Stabilizers in the Substrate

6.1
Introduction

As has been shown, stabilizers have to protect the substrate effectively during processing and in the course of the desired lifetime. For this reason, comprehensive testing of the finished article under the conditions to which the article is exposed in service is the best way to choose the right stabilization system. In the production of complex parts it is important that there is no stress which causes relaxation or post crystallization in service, e.g. outdoor weathering at varying temperatures, leading to cracks that are not the result of oxidative degradation. However, this strongly influences such degradation. Figure 6.1 shows such crack formation in a car bumper, running parallel with the direction of flow after a short service life.

Simulation of the manufacturing process with small substrate and stabilizer quantities is possible. Tests on small parts yield information regarding the efficiency of the stabilizers used and are applicable to the production of large parts. However, the greater shearing forces and residence time in the corresponding machines have to be taken into account. The preparation of test specimens is of great importance and should be carried out only in machines suitable for the purpose. Such tests intended to optimize process stabilizers providing effective protection of the substrate, are not time consuming.

In contrast, finished articles are sometimes exposed to varying influences for years, depending on application. Testing of the changes of the material's properties would require years. Hence, it is not surprising that test methods accelerating aging have been developed. Such methods today represent the state of the art in accelerated aging. In using these tests, it has to be ensured that aging correlates with service conditions. Many of these test methods are recorded in standards (see Appendix 3). Table 6.1 summarizes the most common methods used to determine material properties after processing and aging under long term oxidative conditions.

Foster [320] gave an overview of the analytical methods that can be used in determining material properties during oxidative aging. The

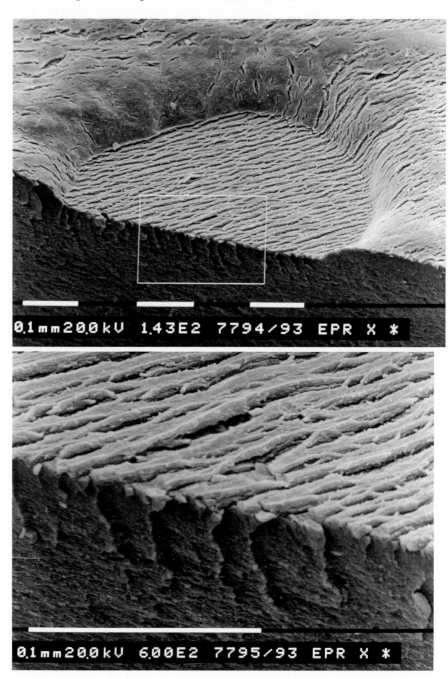

Fig. 6.1. Craze formation caused by relaxation of internal stress in the flow zone

Table 6.1. Methods and material properties according to processing, thermo- and photooxidative aging of polymers

PROCESSING	LONG TERM THERMAL STABILITY	LIGHT STABILITY

Melt Processing
• Melt Mass-Flow Rate [MFR]
• Melt Volume-Flow Rate [MVR]
• Torque Increase
• Viscosity
  ❑ Low Shear / High Shear Viscosity [η]
  ❑ Solution Viscosity
• Moduli
  ❑ Loss Modulus
  ❑ Storage Modulus
  ❑ Polydispersity Index [PDI]
• GPC [M_w, M_n, M_z]
• Color [Y.I.]
	Oven Aging
	• Elongation
	• Tensile Strength
	• Tensile Impact Strength
	• Charpy Impact
	• Izod Impact
	• Time To Embrittlement
	• Color [Y.I.]
	• Haze
	• Gloss
	• Craze Formation
	  ❑ Visual
	  ❑ Microscope
	  ❑ Surface Roughness [RA]
	• Oxygen Uptake
	• Oxygen Induction Time [O.I.T.]
	• Carbonyl Index
	• GPC [M_w, M_n, M_z]
	• Weight Loss [TGA]
		Natural Weathering
		• Europe (e.g. Bandol France)
		• America (e.g. Florida USA)
		Accelerated Natural Weathering
		• Emma
		• Emmaqua
		Artificial Weathering
		• Carbon Arc
		• Xenon Arc
		  ❑ Dry
		  ❑ Dry / Wet Cycle
		• Fluorescent Sun lamp
		
		Tests According To Those Of Long Term Thermal Stability

following chapter deals with factors and relationships deserving special attention regarding accelerated testing.

6.2
Testing of Melt Stability

Prior to processing of a polymer in the melt, the stabilizers to be tested have to be mixed thoroughly with the substrate in order to achieve homogeneous distribution within the substrate. If all components are in powder form, homogeneous blending is particularly easy to achieve. It is also possible to add the stabilizers in a volatile solvent in which the polymer to be stabilized is not soluble. This is done in an inert atmosphere with careful evaporation of the solvent. This method is, however, practical only for the production of small quantities.

The influence of the different stabilizers on the stability of the melt during its processing is determined by measuring some characteristic material properties of the polymer such as molecular weight distribution and discoloration before and after processing. The most informative tests are those that come closest to the processing conditions in practice.

If only relatively small processing facilities are available with correspondingly low shear or very short residence time, more severe processing conditions can be simulated by means of multiple extrusions. Other test methods are based on measuring the change of melt viscosity during processing in a kneader with a closed chamber. In the event of molecular weight decrease, the time to drop in torque of the kneading heads is measured. In the event of molecular weight increase, analogous increase of torque can be observed. Testing the stability of a polymer by multiple extrusions is to be preferred to using a kneader.

Measurement of the molecular weight, M_w and the molecular weight distribution, M_w/M_n, can be carried out by means of gel permeation chromatography, GPC. This method yields good results for polymers that have undergone chain scission during processing. Molecular weight increase caused by chain branching or crosslinking, as can be observed with some polyethylene types, is difficult to determine analytically and erroneous conclusions are often drawn [321].

Because changes of the material's properties caused by processing manifest themselves predominantly in the rheological properties of the polymer and thus influence its processability, rheological measuring methods are increasingly being used to characterize the polymer before and after processing.

The simplest and quite informative method is based on the determination of the melt viscosity. The melt mass-flow rate, MFR, and the melt volume-flow rate, MVR, permit conclusions regarding change in molecular weight and change of molecular weight distribution respectively. The use of a capillary rheometer permits accurate determination of rheological material functions, particularly under the influence of high shearing forces.

Furthermore, it is possible to determine die swell under processing conditions and the critical throughput characteristic for melt breakdown. For the determination of the molecular weight and the molecular weight distribution in the melt, viscometers with plate/plate or cone/plate configurations are particularly suitable.

The viscosity of a melt is determined as a function of the speed of revolution ω or of a shear rate γ'. Determination of storage modulus $G'(\omega)$ and loss modulus $G''(\omega)$ allows the calculation of the complex viscosity η^* which in turn permits conclusions regarding molecular weight at low shear rates. The position of the crossover point for $G'=G''$ indicates changes in molecular mass distribution for a given polymer [321]. The polydispersity index [PDI] is referred to as the inverse value of the crossover frequency [322]. Figure 6.2 depicts the "rheological spectrum" of a polypropylene type as determined with a viscometer having plate/plate configuration.

Along with the determination of the rheological properties of a material, an important test criterion is discoloration of the polymer as a function of processing. Usually the yellowness index, Y.I., is measured,

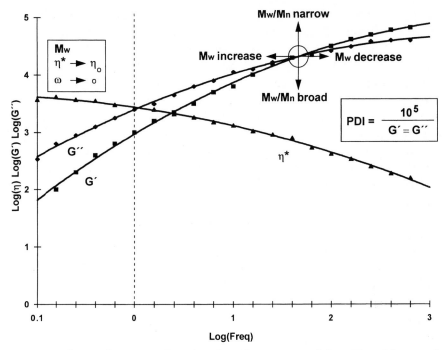

Fig. 6.2. Polypropylene: Complex viscosity, η^*, storage modulus, G', and loss modulus, G'' as a function of γ'. (240 °C, freq. sweep 0.1–500 rad/s, 15% strain)

although measurement of other values such as shading, degree of whiteness, and transparency are possible. Chemical analyses determining stabilizer concentrations before and after processing reveal whether the concentrations used are sufficient for adequate protection during processing in the melt and for the desired lifetime of the end-use article.

6.3
Testing of Long Term Thermal Stability

6.3.1
Oven Aging Techniques

For the determination of the thermooxidative stability of a plastics material, polymer samples are usually subjected to accelerated aging in an oven. The determination of long term thermal stability of polymers by aging in an oven is described in detail by Forsman [323] and Drake [324] and the necessary precautions pointed out, e.g. to avoid cross-contamination by differently stabilized samples in the same oven. To increase the effect of accelerated aging even further in order to obtain faster estimates of a possible lifetime, measurement of the thermooxidative stability at temperatures above the melting point of the polymer has been suggested, and is still being recommended. The thermo-analytical methods DSC and DTA are based on the time necessary to initiate oxidation of the sample, e.g. oxidation induction time, O.I.T. Many investigations concerning this topic can be found in the literature, but Kramer and co-workers [325, 326] have shown, with polyolefins as examples, that values obtained at the high temperatures of the melt do not correlate with data determined below the polymer's melting point. The authors explain this fact with the changes of free volume above the melting point when mobility and solubility of the additives change abruptly. Audoin and co-workers [327] explain this phenomenon with the change of the stabilizer's concentration in the partially crystalline substrate. Below the melting point, the molecular-dispersed stabilizer is "dissolved" only in the amorphous area, while above the polymer's melting point there is a homogeneous solution of the stabilizer in the polymer. Only very few investigations are mentioned in the literature that describe aging behaviour of plastics over a broad temperature range.

Gugumus [328] investigated the aging of polypropylene films at oven aging temperatures between 60 and 149 °C .

Figure 6.3 illustrates the data as an Arrhenius diagram. O.I.T data obtained with polypropylene films with the same stabilization at temperatures above the melting point are also shown [329]. In contrast to the findings of Audouin et al. [327], at temperatures above and below the melting point, not only the pre-exponential factor of the Arrhenius expression changes, but also the corresponding activation energies.

The apparent S-shape of the Arrhenius curve can also be caused by a change of the polymer's morphology expressed as the mean radius of the free volume, measured by positron annihilation, in the range 70–80 °C [330], as well as by physical loss of the stabilizers during prolonged exposure of the polymer.

Gugumus [331] and Pauquet [332] have shown that data such as MFR and long term thermal stability do not correlate with O.I.T. data obtained with the same polymer and the same stabilization. Figure 6.4 depicts a comparison of O.I.T. data obtained at 185 °C and long term thermal aging in an oven (149 °C) for polypropylene samples stabilized in different ways.

It was found that the O.I.T. value increases strongly for stabilization with a phenol/phosphite blend with a given phenol concentration and

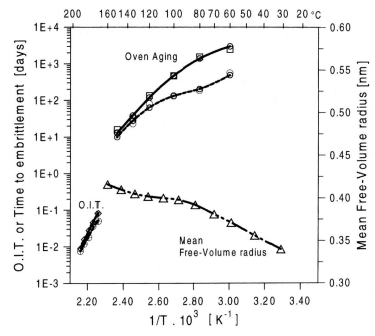

Fig. 6.3. Arrhenius plots for aging of polypropylene above and below its melting point. Samples: 120 μm thick films. ⊕ O.I.T: 0.05% AO-18 + 0.05% PS-2. ◇ O.I.T: 0.05% AO-18 + 0.05% PS-2 + 0.1% HAS-9 + 0.1% HAS-10. ○ Days to Brittleness: 0.05% AO-18 + 0.05% PS-2. □ Days to Brittleness: 0.05% AO-18 + 0.05% PS-2 + 0.1% HAS-9 + 0.1% HAS-10

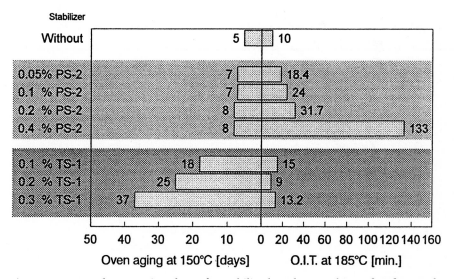

Fig. 6.4. O.I.T. and oven aging data of a stabilized PP homopolymer [329]. Sample thickness: 0.5 mm. All samples contain: 0.05% AO-18 and 0.05% Ca-stearate

increasing concentration of the phosphite, while oven aging at 149 °C and with the same stabilization systems exhibits only minor differences.

Zweifel [333] has shown that during oven aging at 149 °C the phosphite is transformed quantitatively to phosphate in a short time and the life time of the polymer then depends only on the phenol concentration. Exactly the opposite occurs if a phenol/thiosynergist blend is used for stabilization. The O.I.T. value remains practically constant with increasing thiosynergist concentration but oven aging time increases strongly. Prediction of the thermooxidative stability of polymers based on O.I.T. data measured in the melt, therefore, leads generally to erroneous conclusions.

Gugumus [331] has shown that O.I.T. data could be used for analytical purposes for selected stabilizers in the substrate.

The time required to determine the foreseeable lifetime based on oven aging data depends on the thickness of the samples. Gijsman [334] describes this relationship in the case of aging of polypropylene samples of different thickness, stabilized with 0.1% AO-3 as shown in Fig. 6.5.

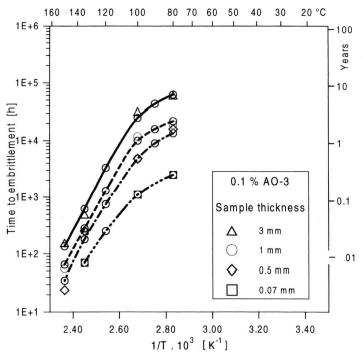

Fig. 6.5. Arrhenius plots for aging of polypropylene samples with different sample thickness

However, it has to be taken into account that within a test series with a given stabilizer, changeover may occur from diffusion to evaporation-controlled loss.

Furthermore, air velocity and flow inside the oven can influence measurements. Aging temperature strongly influences the measurement of mechanical properties such as time to embrittlement on bending, elongation, tensile strength and tensile impact strength of partially crystalline polymers like polyolefins. Annealing of polypropylene at 150 °C leads to dramatic changes of material properties such as elongation, with simultaneous increase in density, melting point and stiffness [335]. These changes are caused by a changeover from smectic to monoclinic crystal form. For this reason, great caution has to be exercised in the choice of test conditions for accelerated aging experiments to avoid arriving at erroneous conclusions with regard to the polymer's behaviour under real service life conditions.

Along with monitoring of the mechanical properties of polymers in the course of aging, it is also possible to measure oxidation products resulting from thermooxidative degradation. Monitoring of the intensity of the $C=O$ absorption in the IR-spectrum provides information regarding the progress of aging.

Values measured across the sample's cross-section, e.g. by microtoming in thin slices, indicate whether oxidation is diffusion-controlled, i.e. whether layers close to the surface are preferentially oxidized, while the sample's interior is relatively intact.

Additional information concerning the material's behaviour after long term thermal aging is provided by measurements of discoloration, surface roughness, micro crack formation and loss of gloss.

6.3.2
Testing Under External Stress

Results presented so far are based on long term thermal aging of samples not subjected to external stress. Rapoport and co-workers [336, 337] have found that the rate of oxidation of oriented polypropylene fibres at 130 °C increases under external stress. Horrocks and co-workers [338] have shown that applying stress in aging experiments reduces time to failure and that the results are influenced by the type of antioxidant used. White and Rapoport [339] have summarized the most important results concerning the effect of stress, pointing out that tensile stresses have predominant effect on aging behaviour. The formation of micro cracks as source of failure under uniaxial tension is described by Zhurkov and co-workers [340]. Rapoport and co-workers [341, 342] point out that crack formation originating at initiating points proceeds faster under extreme stress because of the heterogeneous character of polypropylene oxidation.

Systematic investigations of the influence of stress on thermal aging behaviour of polymers have been carried out mainly in long term tests of water pipes that have been exposed to hydrostatic pressure at temperatures between 60 and 120 °C (depending on the material). Ifwarson and co-workers have conducted such investigations over a period of 20 years mostly using polyolefins [343, 344]. One of the advantages of testing long term thermal aging of pipes is the possibility of measuring under hydrostatic pressure, i.e. externally applied stress. In addition, different internal and external media can be chosen such as water and air. In this way extractability of the stabilizers can also be studied. Tests of the relationship between hydrostatic pressure and time are generally represented as plots of log (hoop stress)/log (failure time) under isothermal conditions.

At high hoop stress ductile break occurs which is also referred to as stage I failure as shown in Fig. 6.6. This behaviour is usually observed with polyolefin pipes.

The changeover from ductile failure to break caused by brittleness occurs when oxidation of the material surface has proceeded so far that under the given pressure crack formation and propagation leads to stage III failure, as shown in Fig. 1.13. In the literature [343] there is also mention of defined areas where ductile, as well as brittle failure (so-called phase II failure) occurs.

Kramer [345] did not observe phase II failure in investigations of polybutene-1 pipes. Similar findings are reported by Dörner and Lang [346] for PE-MD pipes as shown in Fig. 6.7. The time for sudden change-over from ductile to brittle failure depends on the stabilizers used.

Gedde and co-workers [347] and Dörner and Lang [346] have shown that at the point of abrupt changeover, stabilizers have been consumed. This may be the result of consumption leading to non-active components caused by the stabilizers' reaction characteristics, as well as purely physical loss caused by migration.

The Arrhenius relationship permits a rough estimate of the probable life time. Figure 6.8 depicts an Arrhenius plot of the aging of pipes made of PE-MD [344] wihout stabilizer, with AO-21 as phenolic antioxidant having dual functionality and with a blend of the phenolic antioxidant, AO-18, and the phosphite, PS-2.

The kink in the curve at approx. 85 °C could be due to the T_α relaxation of PE-MD [348]. The task group plastic pipes and fittings within the

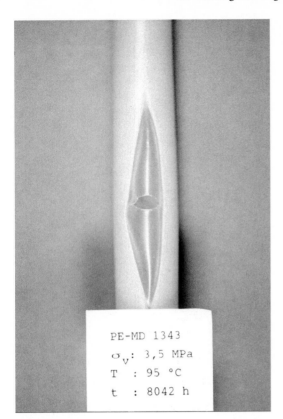

Fig. 6.6. Picture of ductile failure of a PE-MD-pipe

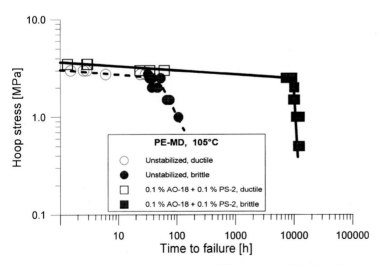

Fig. 6.7. Hoop stress versus time to failure at 105 °C. Sample size: PE-MD, extruded pipe (20 mm diameter, 2 mm wall thickness). G. Dörner, Ph.D. Thesis, Montanuniversität Leoben, Austria, with permission

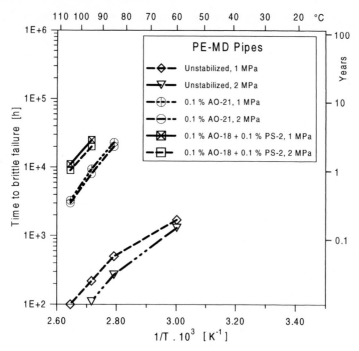

Fig. 6.8. Arrhenius plot of hoop stress experiments involving PE-MD pipes with different stabilizers. G. Dörner, Ph.D. Thesis, Montanuniversität Leoben, Austria, with permission

working groups of ISO/TC 138 has suggested in its draft ISO/DP 9080 [349] a physics-based extrapolation according to Arrhenius. This approach is based on a combination of Eyring's theory (Eq. 5.7) with statistical methods for assessment of the data. It also permits mathematically exact description of internal pressure and time-related status curves. However, this method requires extremely extensive measurements and for this reason is rarely applied in practice.

6.3.3
High Pressure Oxygen Techniques

Accelerated aging of polymers under thermooxidative conditions and increased oxygen pressure has been repeatedly attempted. Gillen and Clough [351] have shown in their overview paper dealing with the problems associated with monitoring the oxidation of polymers that raising oxygen pressure strongly accelerates oxidation.

In using high oxygen pressure, particularly with relatively thick samples, it has to be ensured that measurements are not influenced by change over from a diffusion-controlled oxidation with preferential oxidation of the surface to homogeneous oxidation across the specimen cross-section. Information obtained from oxidation profiling techniques helps in understanding these effects.

6.3.4
Oxygen Absorption Techniques

Determination of oxygen consumption for the duration of the experiment by measuring oxygen uptake was described early as a possible method in assessing the thermooxidative stability of polymer [352]. Hawkins [353] has pointed out that this method has advantages over simple oven aging. This is because the commonly used samples, as a result of reduced thickness (up to 250 μm), do not allow diffusion-controlled oxidation and the static experimental conditions in a closed container do not permit the evaporation of the stabilizers. Horng and Klemchuk [354] have shown that with this method it is possible to determine the probable lifetime of a given polymer, unstabilized as well as a function of the stabilizers used. The time to significant oxygen uptake at 40, 100 and 140 °C was correlated with mechanical measurements such as elongation and carbonyl absorption.

The authors found only insignificant changes of the above-mentioned criteria in the course of the induction period. Wiles and co-workers [355] have shown that oxygen uptake during the induction period is not zero, although it is so small that it couldn't be measured with techniques available so far. Clough and co-workers [356] have found in aging of nitrile rubber that oxygen uptake at 23 °C can be measured with appropriate techniques (rate of oxygen consumption approx. 10^{-13} mol g^{-1} s^{-1}) and that such measurements correlate well with those obtained at higher temperatures.

Assessment of the corresponding Arrhenius plot permits conclusions regarding the probable life time of the material at room temperature.

6.3.5
Chemiluminescence Techniques

More than 30 years ago Ashby [357] described the appearance of chemiluminescence during oxidation of polymers. The reaction causing this phenomenon is the bimolecular reaction of two peroxy radicals forming a keto group in an excited state. This returns to its original state under emission of light. Reich and Stivala [358] have shown that the intensity of the emission is directly proportional to the resulting carbonyl concentration. Billingham and Then [359] have found that by measuring the intensity of the emission of the luminescence it is possible to obtain direct correlation with the content of the formed peroxide groups.

George [360] described the use of chemiluminescence measurements for the study of the kinetics of polymer oxidation. Measurements are reported carried out with a variety of polymers such as polypropylene, polyethylene, polystyrene and polyamides.

Ashby [357] and also Schard and Russel [361] have used measurements of chemiluminescence to determine the efficiency of stabilizers as oxidation inhibitors.

Because of the extremely weak light emission, chemiluminescence can be determined only by sensitive photon-counting methods. Thus, conducting routine measurements was practically impossible. Since the recent development of a sensitive charged coupled device, CCD, camera, the use of this technique has been greatly simplified [362]. Dudler and Lacey [319] have shown that chemiluminescence correlates with oven aging data obtained on aging of polypropylene specimens stabilized with various sterically hindered phenols as shown in Fig. 6.9.

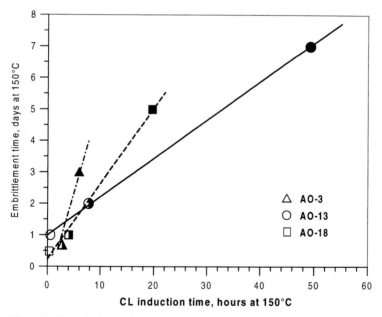

Fig. 6.9. Correlation of CL-induction time with oven embrittlement time for PP films (100 μm), containing various amounts of different stabilizers (control: □, △, ○), 0.05% (◨, ◭, ◑) and 0.1% (■, ▲, ●)

Compared with the determination of aging time in an oven, the chemiluminescence method permits time reduction by a factor of 4 to 12. The acceleration is likely caused by the fact that chemiluminescence measurements are carried out in pure oxygen atmosphere, while aging in the oven is conducted in air. It is known that increasing oxygen pressure accelerates thermal aging [350] (see also Sect. 6.3.3).

Hosoda and co-workers [363] have shown that, by means of chemiluminescence-imaging, it is possible to visualize stress situations in the examined specimens.

The further development of chemiluminescence measurements as a standard method for the determination of the probable lifetime of a polymer needs further experimental and theoretical investigations.

6.4
Testing Light Stability of Polymers

6.4.1
Natural vs Artificial Weathering

Testing plastic articles regarding their stability against photooxidative degradation is of great importance. Practically all plastics are exposed during service to light and/or external weathering conditions. Because such materials should preserve their properties and appearance over years, they are protected by the use of suitable light stabilizers. A test of their stability under normal service conditions can last years. The development of methods permitting acceleration of the measurements was, and still is, considered very important. Such acceleration can be achieved by the choice of a geographically suitable site such as Florida or Phoenix (Arizona) having intensive and long solar radiation. On the other hand aging can be accelerated by "strengthening" of the sunlight by appropriate measures such as the use of mirrors, e.g. EMMA (Equatorial Mount with Mirrors for Acceleration) or EMMAQUA (in addition to EMMA under water spraying). As a result, the intensity of normal sunlight is increased ninefold across the whole emission spectrum. The intensity of light with a wave length of >400 nm is increased, although only threefold [364]. At present, acceleration is generally achieved by using weathering devices with suitable lamps [365]. Davis and Sims [364] comprehensively described the behaviour of various plastics under different weathering conditions. Gugumus [366] has shown that best correlation between natural and artificial weathering is obtained with devices having emission spectra similar to that of sunlight. It is most important to use radiation sources in combination

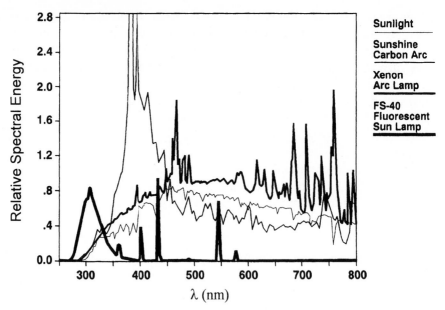

Fig. 6.10. Emission spectra of different light sources compared to sun light – a comparison of relative spectral energy distribution

with suitable filters impermeable to UV radiation below 295 nm. Figure 6.10 depicts a comparison of the distribution of relative intensity of emissions from various light sources vs sunlight.

Gugumus [366] has also pointed out that stabilizers with relatively high migration ability such as low molecular weight sterically hindered amines of the HAS type, can migrate to the surface during the dark periods. Test results obtained under continuous artificial light exposure without dark periods can, therefore, deviate appreciably from those obtained under natural weathering. For this reason, artificial weathering is frequently conducted in devices with light and dark periods that can be regulated and in addition are equipped with water spraying facility. The temperature at which artificial weathering is carried out influences results considerably because thermooxidative degradation is superimposed on the photooxidative degradation. For comparison purposes, Table 6.2 [367] summarizes results obtained with a PP homopolymer stabilized with various light stabilizers exposed in a device equipped with a Xenon-arc burner and those from weathering experiments conducted in Florida.

It is very difficult to correlate in a general way accelerated artificial weathering, even if conducted in suitable devices with weathering under natural conditions. Attempts to determine acceleration factors per-

Table 6.2. Comparison of results obtained by artifical and outdoor exposure of PP-homopolymer. Exposure device: XENO 1200, time T_{50} to 50% retained tensile strength. Outdoor exposure: Florida, 45° South, energy E_{50} to 50% retained tensile strength. All samples contain: 0.1% Ca-stearate, 0.05% AO-18 and 0.05% PS-1. Sample size: tapes, 50 μm sample thickness

Light stabilizer	XENO 1200, b.p. 55°C Time to T_{50}, hours	Florida, 45° South Energy to E_{50}, kJ/cm^2	Correlation T_{50} / E_{50}
Without	400	75	5.33
0.5 % UVA-13	800	167	4.19
0.5 % UVA-2	1380	188	7.34
0.1 % HAS-5	1300	1003	1.29
0.1 % HAS-10	1630	460	3.54

mitting reliable statements concerning the probable lifetime of a plastic part on the basis of accelerated tests have been, and still are, made. However, results available today are unsatisfactory [364, 366].

Still accelerated weathering provides first indications regarding the useful lifetime of a plastics material as a function of the chosen stabilization system. Testing under outdoor weathering conditions is always preferable.

6.4.2
Test Methods

In principle, the same test methods are being used as those for the determination of a polymer's thermooxidative degradation. For the assessment of photooxidative degradation, along with the usual measurements of mechanical properties, roughness of the surface (ISO/DIS 4287/1, see Fig. 1.17) is also frequently measured. This method allows good characterization of oxidation on the polymer's surface and provides, next to microscopic investigations, early information regarding degradation. Correlation between changes of mechanical properties and the formation of oxidation products, as characterized by carbonyl absorption, also allows insight into the progress of photooxidation. Here, too, determination of the oxidation profile across the sample's cross-section is advantageous to obtain insight into the influence of oxygen diffusion [368].

Determination of oxygen uptake as a means of monitoring photooxidation has been described several times in the literature [369, 370]. Because photooxidation starts on the polymer's surface, the fracture of a polymer part under stress can occur by crack propagation [371–

373]. O'Donnell and co-workers [374] have described the determination of average molecular weight, M_w, across the sample cross-section for polystyrene and polypropylene as a function of aging time. Kim and co-workers [375] investigated measurement of the specific fracture energy as a function of accelerated weathering of polycarbonate. Baumhardt-Neto and De Paoli [376] have shown that photooxidation of polypropylene is influenced by external stress.

6.5
Effect of Gamma Radiation on Polymers

Gamma rays are being used for the sterilization of medical articles made of plastics and also for foodstuffs. The formation of alkyl radicals in the polymer matrix caused by gamma rays leads to accelerated oxidation. Carlsson and co-workers investigated oxidation of low density polyethylene [377], PE-LLD, initiated by gamma rays and that of polypropylene [378]. They have shown that the oxidation products generated by γ radiation differ significantly from those formed by photo- and thermooxidation. The generation of high radical concentration by γ rays leads to accelerated aging of the material. Gillen and Clough [379, 380] described a method for predicting the lifetime of polymers by a combination of short term γ radiation and thermal aging data.

With this method too, it is evident that aging is influenced by the diffusion of oxygen into the material after γ irradiation depending on the dose level, the dose rate, and the geometry of the test specimen. Determination of the oxidation profile across the sample cross-section discloses whether diffusion-controlled oxidation has taken place.

Conclusions

The choice of a plastics material for the production of a final article depends primarily on the properties under the foreseen service conditions. Aging under accelerated conditions should provide information on whether the chosen plastics material fulfills requirements regarding processing and service lifetime by the use of suitable stabilizers. It follows that only such tests need be conducted that permit simulation of the final article's future exposure.

CHAPTER 7

Outlook and Trends

7.1
New Substrates

7.1.1
Metallocene-Based Polymers

The development of polymers based on a variety of α-olefins and prepared by polymerization in the presence of metallocene catalysts (known also as single site catalysts, SSC) leads to new substrates for packaging, construction, textiles and further applications. Primarily, metallocenes based on zirconium are used as SSC catalysts.

Practically all olefins and diolefins such as ethylene, propylene, butene-1, pentene-1, hexene-1, and further α-olefins, 1,3-butadiene, 1,5-hexadiene, cyclic olefins and diolefins such as cyclopentene, norbonene, dicyclopentadiene, and also various styrene monomers and (meth)acrylates can be transformed to polymers with exceptional copolymerization and stereocontrol capabilities.

So far, there are no references to be found in the literature dealing with processing and long term stability of such materials. Because of their chemical composition, it can be assumed that thermooxidative and photooxidative stability does not differ substantially from that of the known and thoroughly described stability of classic polyolefins based, e.g., on chromium or Ziegler-Natta catalyst technology. However, the optimal stabilization formulation for each polymer has to be determined with regard to processing and end application, i.e. processing stabilizers, stabilizers for thermal and photo stability. Of particular importance are the solubility and diffusion of the stabilizers in polymers with low density.

7.1.2
Engineering Thermoplastics

Engineering thermoplastics, including high performance thermoplastics, are being processed at increasingly higher temperatures. This requires processing stabilization withstanding such conditions. Currently known processing stabilizers such as phosphites or phosphonites meet these requirements only partially because they show tendency to thermal decomposition at temperatures above 300 °C. Development of stabilizers for use at high processing temperatures will, therefore, be the target for new products in this sector. Suitable C-radical acceptors may bring about useful results in these applications. At any rate, mechanical and oxidative degradation mechanisms would need further in-depth investigations.

Suitable light stabilization of engineering thermoplastics is unavoidable, because such materials are frequently used under conditions that can cause photooxidative and photo-induced degradation (see Sect. 1.7.1). Most of these polymers, e.g. aromatic polyesters and polyamides, polyethersulfones, polysulfides, polycarbonate and other polymers containing aromatic moieties absorb UV light in the long wavelength segment. For this reason, red shifted UV absorbers, RUVA, are particularly suitable for this application. The UV absorbers have to be inherently photo-stable. They have very low volatility at high processing temperatures, with high molar extinction coefficient, and be readily soluble even at high concentrations in the polymer matrix to prevent crystallization in the polymer. Work is currently under way to develop UV absorbers meeting these requirements. Of prime interest are substances such as benzotriazoles and hydroxyphenyl triazines.

7.1.3
Polymer Blends and Alloys

There are practically no limits to the manufacture of new work materials based primarily on heterophased blends. The oxidative stability of such blends or alloys is determined by the component most prone to oxidative degradation. Particular attention has to be paid to the protection of the individual phases because low molecular degradation products such as peroxides, peracids or aldehydes form volatile prodegradants that migrate into the more stable phase, destroying it gradually by oxidation. Generally, the stabilizers used (processing, long term thermal exposure and light stabilizers) are dissolved uniformly in all phases during processing in the polymer melt. Because the polymer melt is a relatively viscous liquid, the stabilizers remain more or less molecularly dispersed in the individual phases while cooling (see Sect. 3.4.4, Fig. 3.19). Distribution corresponding to the solubility of the stabilizers in the individual phases is gradually achieved through diffusion processes. Protection of an easily oxidizable phase against degradation is only achieved by focused insertion of the stabilizer into that phase where it remains. Stabilizers with particularly high solubility in the phase to be protected may offer possible approaches. This is possible by synthesizing suitable stabilizer molecules. By chemical bonding, e.g. polymer analogous reactions of suitable stabilizers or reactive grafting onto the backbone molecule of the phase to be protected, the given stabilizer is perforce there where it is to be active. However, it would appear that a certain mobility of the stabilizer molecules is necessary to impart optimal protection. There are examples in the literature of polymer-bound stabilizers [381]. These concepts have not been confirmed in practice yet.

7.2
Recycling of Polymers

Polymer waste material resulting from processing, e.g. scrap from injection moulding, material from the start-up of equipment, is generally returned for re-use, e.g. by addition to virgin material in relatively low volume for first use. This procedure is known as post-industrial re-use of scrap or primary recycling. This process presents no great difficulties because the producer knows the material and its stabilization.

In contrast to this is the re-use of plastic waste after first use, the so-called post-consumer waste re-use or secondary recycling. The most simple and obvious is the re-use of plastics that have already been in service whose composition, application and information concerning them is known, such as crates of PE-HD or PP, containers, automotive bumpers, battery cases, and housings of electronic appliances, to mention only a few. It should be borne in mind, however, that recycling of such materials should not be undertaken after the material has been damaged too severely by oxidation.

Methods such as determination of the increase in carbonyl absorption and analysis of stabilizers not yet consumed in the used material provide information concerning the condition of the material.

By the addition of suitable stabilizers for processing and long term protection, the "old material" can be used for new applications. There have been recent publications showing that these recyclates exhibit properties close to those of virgin materials [382–385].

For some time public interest has been focused on the large amount of domestic waste of plastic that represents approximately 7 wt% or 20 vol.% of total municipal waste. A variety of approaches aimed at reducing this amount are being discussed and are subjects of legislation dealing with it. Within the framework of this monograph, only mechanical reuse, i.e. recycling of mixed plastics waste, is of importance. The preparation of practically uniform material according to plastic type is desirable but the required sorting facilities have to be improved substantially. As long as this is not achieved, only mixed plastics waste has to be considered. In this case, neither degree of oxidative degradation of the individual materials, nor the available stabilizers and other additives such as fillers and pigments, are known and analysis in this respect is extremely complex, and therefore cannot be realized in practice. The oxidative behaviour of recycled material from such waste is determined to a large extent by the presence of degradation products acting as prodegradants. It follows that such materials do not possess the properties of the original substrates and can, therefore, be considered only for simple applications. By the addition of new suitable stabilizers, the performance of such materials can be improved during processing and their lifetime, although irreversible damage in the course of first use cannot be eliminated.

For this reason, integrated plastics waste management is of great importance taking into account other methods of re-use such as feedstock recycling or combustion with energy recovery [386].

New plastics materials and processing technologies have recently led to acceptance of plastic waste, particularly in the packaging sector. By reducing thickness of the final article while maintaining mechanical properties, the volume of plastic for the manufacture of such articles is reduced. The multiple use of plastic like PET or PE-HD bottles is also possible. In all these instances, corresponding stabilization is unavoidable prior to processing and long term use.

Adequate stabilization of plastic is the basis for processing and long term protection of plastic articles throughout their lifetime. Increasingly more complex stabilization systems will be used, providing "tailor-made" solutions for the application of plastic in view of the different applications.

References

1. Hoffman, A.W., J. Chem. Soc., 13 (1861) 87
   See also: Scott, G in: "Atmospheric Oxidation and Antioxidants", Ed. Scott, G., Elsevier Science Publishers B.V. Amsterdam, Vol. 1 (1993) 1
2. Bolland, J.L. and Gee, G., Trans. Faraday Soc., 42 (1946) 236
3. Bolland, J.L., Quart. Rev. (London), 3 (1949) 1
   See also: Al-Malaika, S., in: "Atmospheric Oxidation and Antioxidants", Ed. Scott, G., Elsevier Science Publishers B.V. Amsterdam Vol. 1 (1993) 45
   Reich, L. and Stivala, S.S., "Autoxidation of Hydrocarbons and Polyolefins, Marcel Dekker Inc., N.Y. (1969) 1
4. Denisov, E.T., in: Handbook of Antioxidants, Bond Dissociation Energies, Rate Constants, Activation Energies and Enthalpies of Reactions, CRC Press, Boca Raton (1995)
5. Howard, J.A. in: "Free Radicals", Ed. Kochi, J.K., Wiley, N.Y., Vol. 2 (1973)
6. Howard, J.A., Schwalm W.J. and Ingold, K.U., Adv. Chem. Ser., 75 (1968) 6
7. Howard, J.A., Ingold, K.U. and Symonds, M.S., Can. J. Chem., 46 (1968) 1017
8. Bolland, J.L., Trans. Faraday Soc., 46 (1950) 358
9. Korcek, S., Chemier, J.H.B., Howard, J.A. and Ingold, K.U., Can. J. Chem., 50 (1972) 2285
10. Leroy, G., Nemba, R.M., Sana, M. and Wilante, C., J. Mol. Structure, 198 (1989) 159
11. Zaikov, G.E., Howard J.A. and Ingold, K.U., Can. J. Chem., 47 (1969) 3017
12. Russell, G.A., J. Am. Chem. Soc., 79 (1957) 3871
13. Perry R.H. and Chilton, C.H., Chemical Engineering Handbook, 5th Ed., McGraw-Hill, N.Y. (1969)
14. Van Amerongen, G.J., J. Polym. Sci., 5 (1950) 307
15. Michaels A.S. and Bixler, H.J., J. Polym. Sci., 50 (1961) 393, 413
16. Peterson, C.M., J. Appl. Polym. Sci., 12 (1968) 2649
17. Muruganadam, N., Koros W.J. and Paul, D.R., J. Polym. Sci., Part B: Polym. Phys. 25 (1987) 1999
18. Zimmermann, J., J. Polym. Sci., 46 (1960) 151
19. Billingham, N.C., in: "Oxidation Inhibition in Organic Materials", Ed. Pospisil, J. and Klemchuck, P., CRC Press, Boca Raton, Vol. II (1990) 249
20. Zweifel, H., Chimia 47 [10] (1993) 390
21. Sohma, J., Colloid Polym. Sci., 270 (1992) 1060
    See also: Scott,G., in: "Atmospheric Oxidation and Antioxidants", Ed. Scott, G., Elsevier Science Publishers B.V. Amsterdam Vol. 2 (1993) 141
22. Yachigo, S.Y., Sasaki, M., Takahashi, Y., Kojima, F., Takada, T. and Okita, T., Polym. Degrad. Stab., 22 (1988) 63
23. Knobloch, G., Angew. Makromol. Chem., 176/177 (1990) 333

162   References

24. Drake, W.O., Pauquet, J.R., Todesco, R.V. and Zweifel, H., Angew. Makromol. Chem., 176/177 (1990) 215
25. Hinsken, H., Moss, S., Pauquet, J.R. and Zweifel, H., Polym. Degrad. Stab., 34 (1991) 279
26. Ying, Q., Zhao, Y. and Liu, Y., Makromol. Chem., 192 (1991) 1041
27. Moss, S. and Zweifel, H., Polym. Degrad. Stab., 25 (1989) 217
28. Dontula, N., Campbell, C.A. and Connelly, R., Polym. Eng. Sci., 33[5] (1993) 271
29. Gilg, B. Ciba-Geigy AG, Additives Division, Technical Documentation (1992)
30. Iring, M., Szesztay, M., Stirling, A. and Tüdös, F., Pure Appl. Chem., A29[10] (1992) 865
31. Krstina, J., Moad, G. and Solomon, D., Eur. Polym. J., 25[7/8] (1989) 767
32. Kern W., and Cherdron, H., Makromol. Chem., 40 (1961) 177
33. Grassie, N., in: "Polymer Handbook", Ed. Brandrup, J. and Immergut, E.H., Wiley, N.Y., 3rd Edition (1989) II/365
34. Gugumus, F. in: "Kunststoff-Additive", Eds. Gächter, R. and Müller, H., Hanser Verlag, München, 3. Ausgabe (1989) 61
35. Gijsman, P., Hennekens, J. and Vincent, J., Polym. Degrad. Stab., 42 (1993) 95
36. Tüdos, F. and Iring, M., Acta Polymerica 39 (1988) 19
37. Billingham, N.C., Prentice P. and Walker, T.J., J. Polym. Sci., Symposium, 57 (1976) 287
38. Knight, J.B., Calvert, P.D. and Billingham, N.C., Polymer, 26 (1985) 1713
39. Roginsky, V.A., in: Developments in Polymer Degradation, Ed. Grassie N., Applied Science Publishers, London, Vol. 5 (1986) 193
40. Celina M. and George, G.A., Polym. Degrad. Stab., 40 (1993) 323
41. Billingham, N.C., in: "Oxidation Inhibition in Organic Materials", Eds. Pospisil, J. and Klemchuck, P., CRC Press, Boca Raton, Vol. II (1990) 253
42. Billingham, N.C., in: "Atmospheric Oxidation and Antioxidants", Ed. Scott, G., Elsevier Science Publishers B.V. Amsterdam Vol. 2 (1993) 224
43. K.T. Gillen and R.L. Clough, Polymer, 33 (1992) 4358
44. Audouin, L., De Bruijn, J.M., Langlois, V. and Verdu, J., J. Mat. Sci., 29 (1994) 569
45. Rapoport, N., Livanova, N., Balogh, L. and Kelen, T., Intern. J. Polymeric Mater., 19 (1993) 101
46. Factor, A., Ligon, W.V. and May, R.J., Macromolecules, 20 (1987) 2462
47. Kuczkowski, J.A., in: "Oxidation Inhibition in Organic Materials", Eds. Pospisil, J. and Klemchuk, P., CRC Press, Boca Raton, Vol. I (1990) 274
48. Photochemistry of Man-Made Polymers, by McKellar, J.F. and Allen, N.S., Applied Science Publishers, London (1979)
49. Polymer Photophysics and Photochemistry, by Guillet, J., Cambridge University Press, Cambridge (1985)
50. Gijsman, P., Hennekens J. and Vincent, J. Polym. Degrad. Stab., 39 (1993) 271
51. J.R.MacCallum, J.R., in: Developments in Polymer Degradation, Ed. Grassie, N., Applied Science Publishers, London, Vol. 1 (1977) 237
52. Gugumus, F., Macromol. Chem., Macromol. Symp. 27 (1989) 25
53. Ogilby, P.R., Dillon, M.P., Kristiansen M. and Clough, R.L., Macromolecules, 25 (1992) 3399
54. Carlsson D.J. and Wiles, D.M., J. Macromol. Sci., Rev. Macromol. Chem., C14[1] 1976) 65
55. Gugumus, F., Polym. Degrad. Stab., 39 (1993) 117

56. Martin, J.T. and Norrish, R.G.W., Proc. Roy. Soc. (London), A220 (1953) 322
57. Lacoste, J., Carlsson, D.J., Falicki, S. and Wiles, D.M., Polym. Degrad. Stab. 34 (1991) 309
58. Photostabilization of Polymers, Principles and Applications, by J.F. Rabek, Elsevier Applied Science, London (1990)
59. Carlsson, D.J., Garton, A. and Wiles, D.M., in: Developments in Polymer Stabilization, Ed. Scott, G., Applied Science Publishers, London, Vol. 1 (1979) 219
60. Carlsson, D.J. and Wiles, D.M., J. Polym. Sci., Polym. Chem. Ed., Vol. 12 (1974) 2217
61. Gugumus, F., Angew. Makromol. Chem., 158/159 (1988) 151
62. Geuskens, G., Baeyens-Volant, D., Delannois, G., Lu-Vinh,Q., Piret, W. and David, C., Eur. Polym. J., 14 (1978) 291
63. Iring, M., Szesztay, M., Stirling, A. and Tüdös, F., Pure Appl. Chem., A29[10] (1992) 865
64. Day, M. and Wiles, D.M., Polym. Sci. Polym. Lett. Ed., 9 (1971) 665
65. Day, M. and Wiles, D.M., J. Appl. Polym. Sci., 16 (1972) 175
66. Day, M. and Wiles, D.M., J. Appl. Polym. Sci., 16 (1972) 202
67. Rivaton, A., Polym. Degrad. Stab., 41 (1993) 283
68. Rivaton, A., Polym. Degrad. Stab., 41 (1993) 297
69. Marcotte, F.B., Campell, D., Cleaveland, J.A. and Turner, J.A., J. Polym. Sci., A1[5] (1967) 7
70. Tang, L., Lemaire J., Sallet, D. and Mery, J.-M., Makromol. Chem., 182 (1981) 3477
71. Karstens, T. and Rossbach, V., Makromol. Chem., 190 (1989) 3033
72. Gardette, J.-L., Sabel, H.-D. and Lemaire, J., Angew. Makromol. Chem., 188 (1991) 113
73. Grassie N., and Roche, S., Makromol. Chem., 112 (1968) 34
74. Allen, N.S. and McKellar, J.F., Polym. Degrad. Stab., 1 (1979) 47
75. Davis, A., Polym. Degrad. Stab., 3 (1981) 187
76. Nishimura, O. and Osawa, Z., Polym. Photochem., 1 (1981), 191
77. Bellus, D., Hrdlovic, P. and Manasek, Z., Polym. Sci. Polym. Lett. Ed., 4, (1966) 1
78. Abbas, K.B., J. Appl. Polym. Sci., Polym. Symp., 35 (1979) 345
79. Andrady, A.L., Searle, N.D. and Crewdson, L.F.E., Polym. Degrad. Stab. 35 (1992) 225
80. Rivaton, A., Sallet, D. and Lemaire, J., Polym. Photochem., 3 (1983) 463
81. Rivaton, A., Sallet, D. and Lemaire, J., Polym. Degrad. Stab.,14 (1986)1
82. Gardette, J.-L. and Lemaire, J., Makromol. Chem., 182 (1981) 2723
83. Gardette, J.-L. and Lemaire, J., Makromol. Chem., 183 (1982) 2415
84. Gardette, J.-L. and Lemaire, J., Polym. Degrad. Stab., 6 (1984) 135
85. Gardette, J.-L. and Lemaire, J., Makromol. Chem., 185 (1984) 467
86. Peeling, J. and Clark, D.T., J. Appl. Polym. Sci., 26 (1981) 3761
87. Pickett, J.E., ACS Symp.Ser., 280 (1985) 313
88. Gesner, B.D. and Kelleher, P.G., J. Appl. Polym. Sci., 12 (1968) 1199
89. Munro, H.S. and Clark, D.T., Polym. Degrad. Stab., 17 (1987) 319
90. Carlsson, D.J., Chan, L.H. and Wiles, D.M., J. Polym. Sci., Polym. Chem. Ed. 16 (1978) 2365
91. Alvino, W.M., J. Appl. Polym. Sci., 15 (1971) 2123
92. Carlsson, D.J. and Wiles, D.M., Macromolecules, 4 (1981) 179
93. Schoolenberg, G.E., J. Mat. Sci., 23 (1988) 1580

94. Scott, G., in "Atmospheric Oxidation and Antioxidants" 1st. Edition, Elsevier, London (1965)

95. Scott, G., in "Atmospheric Oxidation and Antioxidants", Ed. Scott, G., Elsevier Science Publishers B.V. London, Vol I (1993) 121

96. Pospisil, J., in "Developments in Polymer Stabilization", Ed. Scott, G., Applied Science Publishers, London, Vol. 1, (1979) 1

97. Pospisil, J., J. Adv. Polym. Sci., 36 (1980) 69

98. Henman, T.J. in "Developments in Polymer Stabilization", Ed. Scott, G., Applied Science Publishers, London, Vol. 1 (1979) 39

99. Pospisil, J., Polym. Degrad. Stab., 40 (1993) 217

100. Pospisil, J., Polym. Degrad. Stab., 39 (1993) 103

101. Pilar, J., Rotschova, J. and Pospisil, J., Angew. Makromol. Chem., 200 (1992) 147

102. Bickel, A.F. and Kooyman, E.C., J. Chem. Soc., (1956) 2218

103. Kuczkowski, J.A. in "Oxidation Inhibition in Organic Materials", Eds. Pospisil, J and Klemchuk, P., CRC Press, Boca Raton, Vol. I (1990) 267

104. Denisov, E.T., in "Developments in Polymer Stabilization", Ed. Scott, G., Applied Science Publishers, London, Vol. 3 (1980) 2

105. De Jonge, C.R.H.I., and Hope, P., in "Developments in Polymer stabilization", Ed. Scott, G., Applied Science Publishers London, Vol. 3 (1980) 21

106. Bickel, A.F. and Kooyman, E.C., J. Chem. Soc., (1957) 2220

107. Shlyapintokh, V.Ya. and Ivanov, in "Developments in Polymer Stabilization", Ed. Scott, G., Applied Science Publishers Londen, Vol. 5 (1982) 41

108. Drake, W.O., 14th Intern. Conference on "Advances in the Stabilization and Degradation of Polymers", Conference Proceedings, Ed. Patsis, A.V., Lucerne, (1992) 57

109. Bowry, V.W. and Ingold, K.U., J. Am. Chem. Soc., 114 (1992) 4992

110. Beckwith, A.L.J., Bowry, V.W. and Ingold, K.U., J. Am. Chem. Soc., 114 (1992) 4983

111. Denisov, E.T.,in "Developments in Polymer Stabilization", Ed. Scott, G., Applied Science Publishers, Londen, Vol. 5 (1982) 37

112. Denisov, E.T., Polym. Degrad. Stab., 25 (1989) 209

113. Al-Malaika, S., Omikorede, E.O. and Scott, G., J. Appl. Polym. Sci., 33 (1987) 703

114. Felder, B., Schumacher, R. and Sitek, F., Am. Chem. Soc., Symp. Ser., 151 (1981) 65

115. Step, E.N., Turro, J., Gande, M.E. and Klemchuk, P.P., J. Photochem. Photobiol. A: Chem., 74 (1993) 203

116. Step, E.N., Turro, J., Gande M.E. and Klemchuk, P., Macromolecules, 27 (1994) 2529

117. Step, E.N., Turro, J., Gande, M.E. and Klemchuk, P.P., Polym. Preprints, Vol. 34[2] (1993) 231

118. Gijsman, P., Polym. Degrad. Stab., 43 (1994) 171

119. Seltzer R., Ravichandran R. and Patel, A., US 4'876'300, to Ciba-Geigy Corp

120. Chakraborty, K.B. and Scott, G., J. Polym. Sci., Polym. Lett., 22 (1984) 553

121. Yachigo, S., Sasaki, M., Takahashi, Y., Kojima F., Takada, T. and Okita, T., Polym. Degrad. Stab., 22 (1988) 63.)

122. Yachigo, S., Ida, K., Sasaki, M. Inoue, K. and Tanaka, S., Polym. Degrad. Stab., 39 (1993) 317

123. Yachigo, S. Kanako, I. and Saski, M., Polym. Preprints, Vol. 34[2] (1993) 164

124. Hinsken, H., Mayerhöfer, H., Müller, W. and Schneider H.J., US 4'325'863, to Ciba-Geigy Corp
125. Pitteloud, R. and Dubs, P., Chimia, 48, (1994) 417
126. Schwetlick, K., Pure & Appl. Chem., Vol. 55 (1983) 1634
127. Schwetlick, K., König, T., Rüger, C., Pionteck J., and Habicher, W.D., Polym. Degrad. Stab., 15 (1986) 97
128. Al-Malaika, S., in "Atmospheric Oxidation and Antioxidants", Ed. Scott, G., Elsevier Science Publishers B.V., London, Vol I (1993) 161
129. Shelton, J.R. and Davis, K.E., Int. J. Sulfur Chem., 8 (1973) 197
130. Pospisil, J., in "Oxidation Inhibition in Organic Materials", Eds.Pospisil, J. and Klemchuk, P., CRC Press Inc., Boca Raton, Florida, Vol. 1 (1990) 40
131. Drake, W.O., Pauquet, J.R., Zingg J. and Zweifel H., Polym. Preprints, Vol. 34(2) (1993) 174
132. Al-Malaika, S. in "Atmospheric Oxidation and Antioxidants", Ed. Scott, G., Elsevier Science Publishers B.V. London, Vol I (1993) 184
133. Pospisil, J., in "Oxidation Inhibition in Organic Materials", Eds. Pospisil, J. and Klemchuk, P., CRC Press, Boca Raton, Vol. I (1990) 173
134. Pospisil, J., Polym. Degrad. Stab., 39 (1993) 110
135. Hähner, U., Habicher, W.D. and Chmela, S., Polym. Degrad. Stab., 41 (1993), 198
136. Drake, W.O. and Cooper, K.D., SPE Polyolefins VIII Intern. Conference, Conference Proceedings, Houston, (1993) 417
137. Pospisil, J., Polym. Degrad. Stab., 39 (1993) 104
138. Allen, S.N., Hamidi, A., Loeffelman, F.F., Macdonald, P., Rauhut, M. and Susi, P.V., Plast. Rubber Proc. Appl., 5 (1985) 259
139. Vyprachticky, D. and Pospisil J., Polym. Degrad. Stab., 27 (1990) 227
140. Heller, H.J., Eur. Polym. J. Suppl., (1969) 105
141. Heller, H.J. and Blattmann, H.R., Pure & Appl. Chem., 30 (1972) 145
142. Otterstedt, J.E.A., J.Chem.Phys., 58[12] (1973) 5716
143. Kramer, H.E.A., farbe + lack, 92[10] (1986) 919
144. Kramer, H.E.A., 13th Intern.Conference on Advances in the Stabilization and Degradation of Polymers, Conference Proceedings, Ed. Patsis, A.V., Lucerne, (1991) 59
145. Heller, H.J. and Blattmann, H.R., Pure & Appl. Chem., 36 (1973) 141
146. Rieker, J., Lemmert-Schmitt, E., Goeller, G., Roessler, M., Stueber, G.J., Schettler, H., Kramer H.E.A., Stezowski, J.J., Hoier, H., Henkel, S., Schmid, A., Port, H., Wiechmann, M., Rody, J., Rytz, G., Slongo, M. and Birbaum, J.L., J. Phys. Chem., 96 (1992) 10225
147. Chakraborty, K.B. and Scott, G., Eur. Polym. J., 13 (1977) 1007
148. Vink, P. and Van Veen, Th.J., Eur. Polym. J., 14 (1978) 533
149. Hodgeman, D.K.J., J. Polym. Sci., Polym. Lett. Ed., 16 (1978)161
150. Pickett, J.E. and Moore J.E., Polym. Degrad. Stab. 42 (1993) 244
151. Vink, P., in "Developments in Polymer Stabilization", Ed. Scott, G., Applied Science Publishers, London, Vol. 3 (1980) 122
152. Gugumus, F., Polym. Degrad. Stab., 39 (1993) 121
153. Chien, J.C.W. and Connor, W.P., J. Am. Chem. Soc., 90 (1968) 1001
154. Carlsson, D.J. and Wiles, D.M., Macromolecules, 7 (1974) 259
155. Ranaweera, R.P.R. and Scott, G., Eur. Polym. J., 12 (1976) 591
156. Carlsson, D.J., and Wiles, D.M., J. Polym. Sci., Polym. Chem. Ed., 12 (1974) 2217

157. Pospisil, J., in "Developments in Polymer Photochemistry", Ed. Allen, N.S., Applied Science Publishers, Barking, Vol. 2 (1981) 53
158. Pospisil, J., Polym. Degrad. Stab., 40 (1993) 230
159. Gugumus, F., in: Current Trends in Polymer Photochemistry, Eds. Allen, N.S., Edge, M., Bellobono, I.R. and Selli, E., Ellis Horwood, N.Y. (1995) 255
160. Chmela, S., Carlsson, D.J. and Wiles, D.M., Polym. Degrad. Stab. 26 (1989) 185
161. Rytz, G. and Zweifel, H., unpublished results
162. Kletecka, G., SPE – Polyolefins VII, Conference Proceedings, Houston, (1991)
163. Winter, R., Galbo, J.P. and Seltzer R., US Patent 5'204'473, to Ciba-Geigy Corp
164. Valet, A., farbe + lack, 9 (1990) 689
165. Gugumus, F., Angew. Makromol. Chem., 137 (1985) 189
166. Zweifel, H., in: Lifetime Degradation and Stability of Macromolecular Materials, Advances in Chemistry Series No. 249, Eds. Clough, R.L, Gillen, K.T. and Billingham, N.C., American Chemical Society, Washington, D.C. (1996) 375
167. Haruna, T., 16th Intern.Conference on Advances in the Stabilization and Degradation of Polymers, Conference Proceedings, Ed. Patsis, A.V., Lucerne, (1994) 129
168. Moss, S., Pauquet, J.-R. and Zweifel, H., 13th Intern. Conference on Advances in the Stabilization and Degradation of Polymers, Conference Proceedings, Ed. Patsis, A.V., Lucerne, (1991) 203
169. Todesco, R.V., Polyethylene World Congress SP '92, MAACK Conference Proceedings, Zurich, (1992)
170. Glass, R.D. and Valange, B.M., Polym. Degrad. Stab., 20 (1988) 355
171. Gugumus, F., in Kunststoff-Additive, Eds. Gächter, R. and Müller, H., Hanser Verlag, München, 3.Auflage, (1989) 57
172. Gugumus, F., Polym. Degrad. Stab., 24 (1989) 292
173. Drake, W.O., 14th Intern. Conference on Advances in the Stabilization and Degradation of Polymers, Conference Proceedings, Ed. Patsis, A.V., Lucerne, (1992) 57
174. Gugumus, F., Polym. Degrad. Stab., 44 (1994) 290
175. Gugumus, F., Polym. Degrad. Stab., 44 (1994) 299
176. Todesco, R.V., Polypropylene '94, MAACK Conference Proceedings, Zurich, (1994)
177. Gijsman, P., Hennekens, J. and Vincent, J., Polym. Degrad. Stab., 42 (1993) 101
178. Smeltz, H.R. and Krucker, W., Textilveredlung, 20[9] (1985) 272
179. Klemchuk, P.P. and Horng, P.-L., Polym. Degrad. Stab., 34 (1991) 333
180. Zingg, J., Ciba-Geigy, unpublished results
181. Drake, W.O. and Cooper K., RETEC '93 Conference Proceedings, Houston, (1993)
182. Gugumus, F., in: Developments in Polymer Stabilisation, Ed. Scott, G., Elsevier Applied Science Publishers, London, Vol. 8 (1987) 239
183. Gugumus, F., Polym. Degrad. Stab., 39 (1993) 128
184. Gugumus, F., Polym. Degrad. Stab., 44 (1994) 276
185. Pauquet, J.-R., to be published
186. Todesco, R.V., Diemunsch, R. and Franz, T., Technische Textilien, 36, October (1993) T197
187. Meyer, W.W. and Zweifel, H., EURETEC '88, Conference Preprints, Paris, (1988)

188. Zweifel, H., Polyolefins VI, SPE Conference Preprints, Houston, (1986)
189. Meier, HR., Dubs, P., Künzi, HP, Martin, R., Knobloch G., Berttermann, H.,Thuet, B., Borer, A., Kolczach U. and Rist, G., Polym. Degrad. Stab., 49 (1995) 1
190. Gugumus, F. in Kunststoff-Additive, Eds. Gächter, R. and Müller, H., Hanser Verlag, München, 3. Auflage (1989) 76
191. Gilg, B., Ciba-Geigy AG, Basel, unpublished data
192. Mathis, R.D., Kitchen A.G. and Szalla, F.J., US Patent 4'956'408, to Phillips Petroleum Company, USA
193. Hirai, T., Japn. Plastics, October (1970)
194. Kulich, D.M., Wolkowicz, M.D. and Wozny, J.C., Makromol. Chem., Macromol.Symp. 70/71 (1993) 416
195. Gugumus, F. in Oxidation Inhibition in Organic Materials, Eds. Pospisil, J. and Klemchuk, P., CRC Press, Boca Raton, Vol. II (1990), 104
196. Schmitter, A., Ciba Technical Documentation (1994)
197. Gugumus, F., Plastics Additives, Eds. Gächter, R. and Müller, H., Hanser Publishers, Munich, 3rd Edition (1990) 93
198. Gijsman, P., Tummers, D. and Janssen, K., Polym. Degrad. Stab., 49 (1995) 121
199. Janssen, K., Gijsman, P. and Tummers, D., Polym. Degrad. Stab., 49 (1995) 127
200. Schmitter, A., Ciba Technical Documentation (1992)
201. Gugumus, F., in Plastics Additives, Eds. Gächter, R. and Müller, H., Hanser Publishers, Munich, 3rd Edition (1990) 83
202. Botkin, J.H. and Leggio, A., SPE Conference Proceedings, ANTEC, Detroit, (1992)
203. Schmitter, A., Ciba Technical Documentation (1992)
204. Gugumus, F., in Plastics Additives, Eds. Gächter, R. and Müller, H., Hanser Publishers, Munich, 3rd Edition (1990) 255
205. Stohler, F.R. and Berger, K., Angew. Makromol. Chem., 176/177 (1990) 327
206. Didina, L.A., Karmilova, L.V., Tryapitsyna, E.N. and Enikolopian, N.S., J. Polym. Sci., Polym. Lett. Ed., 16 (1967) 2277
207. Pryde, C.A. and Hellmann, M.Y., J. Appl. Polym. Sci., 25 (1980) 2573
208. Kim, A., Bosnyak, C.P. and Chudnovsky, A., J. Appl. Polym. Sci., 51 (1994) 1841
209. Paul, W., Buysch, H.-J., Nising, W. and Scholl, T., DE Pat. Appl. 3617987 A1, to Bayer AG
210. Hähnsen, H., Nising, W., Scholl, T., Buysch H.J. and Grigo, U., EP 0 320 632, to Bayer AG
211. Goossens, J.C., Factor, A. and Miranda, P.M., US 4 861 664, to General Electric Company
212. Henning, J., Angew. Makromol. Chem., 137 (1985) 57
213. Olson, D.R. and Webb, K.K., Macromolecules, 23 (1990) 3762
214. Sorato-Attinger, C. and Schmitter A., submitted for publication
215. Gugumus, F., in Plastics Additives, Eds. Gächter, R. and Müller, H., Hanser Publishers, Munich, 3rd Edition (1990) 246
216. Michaelis, P., Ciba Technical Documentation (1992)
217. Schmitter, A., Ciba Technical Documentation (1994)
218. Rekers, J.W. and Scott, G., US Pat. 4 743 657
219. Scott, G., Al-Malaika, S. and Ibrahim, A., US Pat. 4 959 410

168   References

220. Al-Malaika, S., Ibrahim, A., Rao, M.J. and Scott, G., J. Appl. Polym. Sci., 44 (1992) 1287
221. Hahnfeld, J.L. and Devore, D.D., Polym. Degrad. Stab., 39 (1993) 241
222. Clauss, M., Gilg, B., Schmitter, A. and Stauffer, W., ECM 4th Intern. Conference Additives '95, Conference Proceedings, Clearwater Beach, Florida, (1995)
223. Gaines, G.L., Polym. Degrad. Stab. 27 (1990) 13
224. Downing, F.B., Clarkson, R.G. and Pedersen, C.J., Oil Gas J., 38(II) (1939) 97
225. Hansen, R.H., DeBenedictis, T., Martin, W.M. and Pascale, J.V., J. Polym. Sci. Part A, 2 (1964) 587
226. Hawkins, W.L., Chan, M.G. and Link, G.L., Polym. Eng. Sci., 11[5] (1971) 377
227. Allara, D.L. and Chan, M.G.,J. Polym. Sci. Polym. Chem. Ed., 14 (1976) 1857
228. Allara, D.L. and White, C.W., in Stabilization and Degradation of Polymers, Eds. Allara, D.L. and Hawkins, W.L., Adv.Chem.Ser. 169, American Chemical Society, Washington,D.C. (1978) 273
229. Chan, M.G. and Allara, D.L., Polym. Eng. Sci., 14[1] (1974) 12
230. Sack, S., Schär, S., Steger, E. and Wagner, H., Polym. Degrad. Stab., 7 (1984) 193
231. Wagner, H., Sack, S. and Steger, E., Acta Polym., 34 (1983) 65
232. Chan, M.G., in Oxidation Inhibition in Organic Materials, Eds. Pospisil, J. and Klemchuk, P., CRC Press Inc., Boca Raton, Florida, Vol. I (1989), 228
233. Pryde, C.A. and Chan, M.G., Soc. Plast. Eng. Tech. Pap., 26 (1980) 180
234. von Gentzkow, W., Knapek, E. and Dietrich, I., Z. Naturforsch, Teil A, 41 (1986) 653
235. Müller, H., in Plastics Additives, Eds. Gächter, R. and Müller, H., Hanser Publisher, Munich, (1990), 105
236. Chan, M.G. and Powers, R.A., Soc. Plast. Eng. Pap., 21 (1975) 292
237. Chan, M.G., Kuck, V.J., Schilling, F.C., Dye, K.D. and Loan, L.D., 13th Int.Conference on Advances in the Stabilization and Degradation of Polymers, Conference Proceedings, Ed. Patsis, A.V., Lucerne, (1991), 11
238. Ciba Ltd, Technical Documentation
239. Klingert, B., in: Polymer Additives Product and Market Developments, Conference Proceedings, Chicago (1995)
240. Fay, J.J., Polyolefins IX International Conference, RETEC Conference Proceedings, Houston (1995)
241. Turely, R.S. and Strong, A.B., J. Advan. Mater., (April, 1994) 56
242. Fay, J.J. and King, R.E., in: Geosynthetic Resins, Formulations and Manufacturing, Conference Proceedings, GRI, Drexel University, Philadelphia, December (1994)
243. Gilg, R., in: Fillers and Additives for Thermoplastics and Rubber, Conference Proceedings, Intertech Conferences, Berlin (1994)
244. Rotschova, J. and Pospisil, J., Angew. Makromol. Chem., 209 (1993) 189
245. Drake, W.O., in: Masterbatch 1987, Conference Proceedings, London, (1987)
246. Herrmann, E. and Damm, W., in Plastics Additives, Eds. Gächter, R, and Müller, H., 3rd Edition, Hanser Publisher, Munich, (1990) 637
247. Pauquet, J.R., Ciba, unpublished data
248. Hinsken, H. and Meyer, F., 4th International Conference on Polyproylene Fibres and Textiles, Conference Proceedings, Nottingham, UK, (1990)
249. Todesco, R.V., Diemunsch, R. and Franz, T., Technische Textilien, 36 (1993) 198

250. Zweifel, H., unpublished results
251. Horsey, D., Leggio, A. and Reiniker, R., in: Effects in Plastics, SPE RETEC, Oakbrook, IL., (1994) 209
252. Frank, H.P. and Lehner, H., J. Polym. Sci., Polym. Symp., 31 (1970) 193
253. Billingham, N.C. and Calvert, P.D., Dev. Polymer.Char., 3 (1982) 229
254. Gee, G., Adv. Colloid. Sci., 2 (1946) 145
255. Roe, R.-J., Bair, H.E. and Gieniewski, J. Appl. Polym. Sci., 18 (1974) 843
256. Billingham, N.C. and Calvert, P.D., Dev. Polymer. Deg., 3 (1980) 139
257. Billingham, N.C., Calvert, P.D. and Manke, S.A., J. Appl. Polym. Sci., 26 (1981) 3544
258. Billingham, N.C., Calvert, P.D., Okopi, I.W. and Uzuner, A.,Polym. Degrad. Stab., 31 (1991) 23
259. Flory, P.J., in: Principles of Polymer Chemistry, Cornell, N.Y. 6th Ed. (1967), 508
260. Földes, E. and Turcsanyi, B., J. Appl. Polym. Sci., 46 (1992) 507
261. Schlotter, N.E. and Furlan, P.Y., Polymer, 33 (1992) 3328
262. Shlyapnikova, I.A., Mar'in, A.P., Zaikov, G.E. and Shlyapnikov, Yu.A., Polym. Sci. USSR, 27 (1985) 1948
263. Mar'in, A.P., Shlyapnikova, I.A., Zaikov, G.E. and Shlyapnikov, Yu.A., Polym. Degrad. Stab., 31 (1991) 61
264. Shlyapnikov, Yu.A. and Mar'in, A.P., Eur. Polym. J., 23 (1987) 629
265. Billingham, N.C., in: Atmospheric Oxidation and Antioxidants, Ed. Scott, G., Elsevier, London, Vol. II (1993) 244
266. Moisan, J.Y., Eur. Polym. J., 16 (1980) 979
267. Moisan, J.Y., in: Polymer Permeability, Ed. J. Comyn, Elsevier, London, (1985) 119
268. Billingham, N.C., in: Oxidation Inhibition in Organic Materials, Ed. Pospisil, J. and Klemchuk, P., CRC Press, Boca Raton (1990) 269
269 Dudler, V. and Muinos, C., in: Lifetime Degradation and Stability of Macromolecular Materials, Advances in Chemistry Series No. 249, Ed. R.L. Clough, K.T. Gillen and N.C. Billingham, American Chemical Society, Washington, D.C. (1996) 441
270. Malik, J., Hrivik, A. and Tomova, E., Polym. Degrad. Stab., 35 (1992) 61
271. Malik, J, Hrivik, A. and Alexyova, D. Polym. Degrad. Stab., 35 (1992) 125
272. Dudler,V., to be published
273. Müller-Plathe, F., Acta Polymer., 45 (1994) 259
274. Calvert, P.D. and Billingham, N.C., J. Appl. Polym. Sci., 24 (1979) 357
275. Billingham, N.C., Makromol. Chem., Macromol. Symp., 27 (1989) 194
276. Crank, J., in The Mathematics of Diffusion, Clarendon Press, Oxford, 2nd. Ed (1975)
277. Figge, K. and Rudolph, F., Angew. Makromol. Chem., 78 (1979) 157
278. Billingham, N.C. and Calvert, P.D., in: Developments in Polymer Stabilization, Ed. Scott, G., Applied Science Publishers, Vol. 3 (1980) 173
279. Luston, J., in Developments in Polymer Stabilization, Ed. Scott, G., Applied Science Publishers, Vol. 2 (1980) 185
280. Flynn, J.H., Polymer, 23 (1982) 1325
281. Moisan, J.Y., Eur. Polym. J., 17 (1981) 857
282. Möller, K. and Gevert, T., J. Appl. Polym. Sci., 51 (1994) 899
283. Billingham, N.C., in Atmospheric Oxidation and Antioxidants, Ed. Scott, G., Elsevier, London, Vol II (1993) 276

284. Földes, E., J. Appl. Polym. Sci., 48 (1993) 1910
285. Földes, E., J. Appl. Polym. Sci., 51 (1994) 1586
286. Billingham, N.C., in: Atmospheric Oxidation and Antioxidants, Ed. G. Scott, Elsevier, London, Vol. II (1993) 288
287. Pospisil, J., in: Advances in Polymer Science, Springer-Verlag, Berlin, Vol. 101 (1991) 65
288. Scott, G., in Atmospheric Oxidation and Antioxidants, Ed. Scott, G., Elsevier, London, Vol. II (1993) 279
289. Gugumus, F., Research Disclosure, (September 1981) 357
290. Minagawa, M., Polym. Degrad. Stab., 25 (1989) 134
291. Malik, J., Tuan, D.Q. and Spirk, E., Polym. Degrad. Stab., 47 (1995) 1
292. White, J.R. and Turnbull, A., J. Mat. Sci., 29 (1994) 606
293. McMahon, W., Birdsall, H.A., Johnson, G.R. and Camilli, C.T., J. Chem. Eng. Data, 4 [1] (1959) 57
294. Ghorbel, I., Thominette, F., Spiteri, P. and Verdu, J., J. Appl. Polym. Sci., 55 (1995) 163
295. Ghorbel, I., Akele, N., Thominette, F., Spiteri, P. and Verdu, J., J. Appl. Polym. Sci., 55 (1995) 173
296. Golovoy, A. and Zinbo, M., Polym. Eng. Sci., 29 (1989) 1733
297. Shlyapnikov, Yu.A., in Developments in Polymer Stabilisation, Ed. Scott, G., Applied Science Publishers, London, Vol. 5 (1982) 1
298. Gugumus, F., Polym. Degrad. Stab. 46 (1994) 123
299. Malek, K.A.B. and Stevenson, A., J. nat. Rubb. Res., 7[2] (1992)126
300. Kondo, H., Tanaka, T., Masuda, T. and Nakajima, A., Pure & Appl. Chem., Vol. 64[12] (1992) 1945
301. Raab, M., International Polymer Science and Technology, 21 [6] (1994) T/47
302. White, J.R. and Rapoport, N.Y., TRIP, Vol. 2 [6] (1994) 197
303. Zhurkov, S.N., Zakrevskyi, V.A., Korsukov, V.E. and Kuksenko, V.S., J. Polym. Sci., Part A-2, 10 (1972) 1509
304. Rapoport, N.Y., Livanova, N. Balogh, L. and Kelen, T., Inter. J. Polymer, Mat., 19 (1993) 101
305. Miner, M.A., J. Appl. Mechan., (1945) A-159
306. Shah, C.S., Patni, M.J. and Pamdya, M.V., Polymer Testing, 13 (1994) 295
307. Audouin, L. and Verdu, J., Polym. Degrad. Stab., 31 (1991), 335
308. David, P.K., IEEE Transactions on Electrical Insulation, Vol. EI-22 [3] (1987) 229
309. Chan, M.G., Gilroy, H.M., Johnson, L. and Martin, W.M., 27th International Cable and Wire Symposium (1978), 99
310. Gillen, K.T. and Clough, R.L., Polym. Preprints, 34[2] (1993) 185
311. Gillen, K.T. and Clough, R.L., Polym. Degrad. Stab., 24 (1989) 139
312. Crine, J.-P., Polym. Preprints, 34[2] (1993) 189
313. Crine, J.-P., J. Macromol. Sci. Phys., B 23[2] (1984) 201
314. Crine, J.-P., IEEE Transactions on Electrical Insulation, Vol. EI-22[2] (1987) 169
315. Verdu, J., J.M.S. Pure Appl. Chem., A31[10] (1994) 1389
316. Gugumus, F., Polym. Degrad. Stab., 44 (1994) 273
317. Gugumus, F., Angew. Makromol. Chem., 137 (1985) 201
318. Langlois, V., Audouin L. and Verdu, J., Polym. Degrad. Stab., 40 (1993) 399

319. Dudler, V. and Lacey, D.J., 17. Intern.Conference on Advances in the Stabilization and Degradation of Polymers, Conference Proceedings, Ed. Patsis, A.V., Lucerne, (1995) 57

320. Foster, N.G., in: Oxidation Inhibition in Organic Materials, Eds. Pospisil J. and Klemchuk, P., CRC Press Inc., Boca Raton, Vol. II (1989) 299

321. Hinsken, H., Moss, S., Pauquet, J.-R. and Zweifel, H., Polym. Degrad. Stab., 34 (1991) 289

322. Zeichner, G.R. and Patel, P.D., in: A Comprehensive Evaluation of Polypropylene Melt Rheology, Hercules Inc., Wilmington, DE 19899, USA, (1982)

323. Forsman, J.P., SPE Tech. Papers, 10 (1964) VIII-2

324. Drake, W.O., J. Polym. Sci.: Symposium No. 57 (1976) 153

325. Kramer, E., Koppelmann, J. and Dobrowsky, J., J. Thermal. Anal., 35 (1989) 443

326. Kramer, E. and Koppelmann, J., J. Polym. Sci. Eng., 37 (1987) 945

327. Audouin L., Langlois, V. and Verdu, J., 16th Intern. Conference on Advances in the Stabilization and Degradation of Polymers, Conference Proceedings, Ed. Patsis, A.V., Lucerne, (1994) 11

328. Gugumus, F., AddCon '95, Basel (1995), Conference Proceedings, Rapra Technology Ltd., Shawbury, Shrewsbury, Shropshire SY4 4NR, UK

329. Pauquet, J.-R., Zingg, J. and Zweifel, H., to be published

330. Ohko, Y., Uedono, A. and Ujihira, Y., J. Polym. Sci. Part B: Polym. Phys., 33 (1995) 1190

331. Gugumus, F., in: Developments in Polymer Stabilization, Ed. Scott, G., Vol. 8 (1988) 241

332. Pauquet, J.-R., 42nd Intern. Cable & Wire Symp., Conference Proceedings St.Louis, USA, (1993)

333. Zweifel, H., 13th Intern. Conference on Advances in the Stabilization and Degradation of Polymers, Conference Proceedings, Ed. Patsis, A.V., Lucerne, (1991) 211

334. Gjismann, P., Ph.D. Thesis, Technische Universiteit Eindhoven, NL (1994) 75

335. Drake, W.O., 14th Intern. Conference on Advances in the Stabilization and Degradation of Polymers, Conference Proceedings, Ed. Patsis, A.V., Lucerne, (1992) 59

336. Rapoport, N.Ya., Livanova, N.M. and Miller, V.B., Vysokomol. Soyed., A18 (1976) 2045

337. Livanova, N.M., Rapoport, N.Ya., Miller, V.B. and Musayelyan, I.N., Vysokomol. Soyed., A18 (1976) 2260

338. Horrocks, A.R., Valinejad, K. and Crighton, J.S., Polym. Degrad. Stab., 43 (1994) 81

339. White, J.R., and Rapoport, N.Ya., TRIP, Vol. 2(6) (1994) 197

340. Zhurkov, S.N., Zakrevskyi, V.A., Korsukov, V.S. and Kuksenko, A.F., J. Polym. Sci.: Part A-2, 10 (1972) 1511

341. Rapoport, N.Ya., Livanova, N., Balogh L. and Kelen, T., Intern. J. Polymer. Mater., 19 (1993) 101

342. Rapoport, N.Ya., 17th Intern.Conference on Advances in the Stabilization and Degradation of Polymers, Conference Proceedings, A. Patsis, Ed., Lucerne, (1995) 245

343. Gedde, U.W., Viebke, J., Leijström, H. and Ifwarson, M., Polym. Eng. Sci., Vol. 34, [24] (1994) 1773

344. Gebler, H., Kunststoffe, 79(9) (1989) 823

345. Kramer, E., Ph.D. Thesis, Montanuniversität Leoben (1987)
346. Dörner, G. and Lang, R., submitted for publication
347. Smith, G.D., Karlsson, K. and Gedde, U.W., Polym. Eng. Sci., Vol. 32,[10] (1992) 659
348. Schwarzl, F.R., in: Polymermechanik – Struktur und mechanisches Verhalten von Polymeren, Springer-Verlag, Berlin, (1990) 215
349. ISO standard extrapolation method. Draft proposal ISO/DP 9080.2. ISO/TC 138 N1081 (1987)
350. Faulkner, D.L., J. Appl. Polym. Sci., 31 (1986) 2129
351. Gillen, K.T. and Clough, R.L., in: Handbook of Polymer Science and Technology, Ed. Cheremisinoff, P., Marcel Dekker, New York, Vol. 2 (1989) 167
352. Shelton, J.R. and Winn, H., Ind. Eng. Chem., 45 (1953) 2080
353. Hawkins, W.L., in: Polymer Degradation and Stabilization, Springer Verlag, New York, (1984) 98
354. Horng, P.-L and Klemchuk P.P., in:Polymer Stabilization and Degradation, ACS Symp. Ser. No. 280, Ed. Klemchuk, P., American Chemical Society„ Washington D.C., (1985) 235
355. Grattan, D.W., Carlsson, D.J. and Wiles D.M., Chem. Ind., (April, 1978) 228
356. Wise, J., Gillen, K.T. and Clough, R.L., Polym. Degrad. Stab., in press
357. Ashby, G.E., J. Polym. Sci., 50 (1961) 99
358. Reich, L. and Stivala, S.S., in: Autoxidation of Hydrocarbons and Polyolefins, Marcel Dekker, New York, (1969) Chap. 5
359. Billingham, N.C. and Then, E.T.H., Polym. Degrad. Stab., 34 (1991) 263
360. George, G.A., in: Developments in Polymer Degradation, Ed. Grassie, N., Applied Science Publishers, London, Vol. 3 (1981) 173
361. Schard, M.P. and Russel, C.A., J. Appl. Polym. Sci., 8 (1964) 997
362. Celina, M., George, G.A., Lacey, D.J. and Billingham, N.C., Polym. Degrad. Stab. 47 (1995) 311
363. Hosoda, S., Seki, Y. and Kihara, H., Polymer, 34 (1993) 4602
364. Davis, A. and Sims, D., in: Weathering of Polymers, Applied Science Publishers, London (1983)
365. Brown, R.P., Polymer Testing 12 (1993) 459
366. Gugumus, F., in: Developments in Polymer Stabilisation, Ed. Scott, G., Elsevier Applied Science Publishers, London, Vol. 8 (1987) 260
367. Gugumus, F., Polym. Degrad. Stab., 44 (1994) 276
368. Furneaux, G.C., Ledbury, K.J. and Davis, A., Polym. Degrad. Stab., 3 (1981) 431
369. Grattan, D.W., Carlsson, D.J. and Wiles, D.M., Chem. and Ind., (April 1978) 228
370. Gijsman, P., Ph.D. Thesis, Tech. University Eindhoven, NL, (1994) 143–171
371. So, P.K. and Broutman, L.J., Polym. Eng. Sci., 26 (1986) 1173
372. Schoolenberg, G.E., J. Mat. Sci., 23 (1988) 1580
373. Qayyum, M.M. and White, J.R., Polym. Degrad. Stab. 41 (1993) 163
374. O'Donnell, B., White, J.R. and Holding, S.R., J. Appl. Polym. Sci., 52 (1994) 1607
375. Kim, A., Bosnyak, C.P. and Chudnovsky, A., J. Appl. Polym. Sci., 51 (1994) 1841
376. Baumhardt-Neto, R. and De Paoli, M.-A., Polym. Degrad. Stab., 40 (1993) 53
377. Lacoste, J. and Carlsson, D.J., J. Polym. Sci., Part A: Polym.Chem., 30 (1992) 493

378. Lacoste, J., Vaillant, D. and Carlsson, D.J., J. Polym. Sci., Part A, Polym. Chem., 31 (1993) 715
379. Gillen, K.T. and Clough, R.L., Polym. Degrad. Stab., 24 (1989) 137
380. Gillen, K.T. and Clough, R.L., in: Irradiations Effects on Polymers, Eds. Clegg, D.W. and Collyer A.A., Elsevier Appl. Sci. (1991), 158
381. Pospisil, J., in: Advances in Polymer Science, Springer-Verlag, Berlin, Vol. 101 (1991) 114
382. Drake, W.O., Franz, T., Hoffmann, P. and Sitek, F.A., RECYCLE'91, Conference Proceedings, Davos, (1991) D-1.1
383. Sitek, F.A., Herbst, H., Hoffmann, K. and Pfaendner, R., RECYCLE'94, Conference Proceedings, Davos, (1994) 4–1.1
384. Sitek, F.A., Modern Plast. Int., 10 (1993) 74
385. Pospisil, J., Sitek, F.A. and Pfaendner R., Polym. Degrad. Stab., 48 (1995) 351
386. Mader, F., World Plastic & Rubber Technology (1996), 61

Appendix 1

List of Abbreviations and Symbols

PA	Polyamid
PB	Polybutene-1
PBT	Poly(butylene terephthalate)
PET	Poly(ethylene terephthalate)
PC	Polycarbonate
PMMA	Poly(methyl methacrylate)
POM	Polyacetal/Polyoxymethylene
PE	Polyethylene
PE-HD	Polyethylene, high density
PE-LD	Polyethylene, low density
PE-LLD	Polyethylene, linear low density
PE-MD	Polyethylene, medium density
PP	Polypropylene
TPO	Thermoplastic polyolefin
PS	Polystyrene
BPA-PC	Bisphenol-A based polycarbonate
HIPS	High impact modified polystyrene
MBS	Methylmethacrylate/butadiene/styrene copolymer
SAN	Styrene/acrylonitrile copolymer
PVB	Poly(vinly butyral)
PVC	Poly(vinyl chloride)
ABS	Acrylonitrile/butadiene/styrene copolymer
ASA	Acrylate/styrene/acrylonitrile copolymer
EPDM	Ethylene/propylene/diene copolymer
BR	Butadiene rubber
SBR	Styrene/butadiene rubber
SIS	Styrene/isoprene/styrene block copolymer
NBR	Acrylonitrile/butadiene elastomer
NR	Natural rubber (isoprene rubber)
EPR	Ethylene/propylene rubber
PUR	Polyurethane
T-PUR	Thermoplastic polyurethane

IR	Infrared
UV	Ultraviolet
CL	Chemiluminescence
GPC	Gel permeation chromatography
O.I.T.	Oxidative induction time
TOL	Thickness of oxidized layer
SEM	Scanning electron microscopy
CT-complex	Charge transfer complex
ESIPT	Excited state intramolecular proton transfer
ISC	Inter-system crossing
W.O.M.	Weather.O.Meter
XENO-1200	Exposure device equipped with Xenon-arc burner
b.p. temp.	Black panel temperature
r.h.	Relative humidity
EMMA	Equatorial mount with mirrors for acceleration
DSC	Differential scanning calorimetry
LTTS	Long-term thermooxidative stability
AO	Antioxidant
PS	Processing stabilizer
TS	Thiosynergist
UVA	UV absorber
Q	Quencher
HAS/HALS	Hindered amine stabilizer/Hindered amine light stabilizer
MD	Metal deactivator
HD	Hydroperoxide deactivator
FD	Filler deactivator
CB-A	Chain-breaking acceptor
CB-D	Chain-breaking donor
ISO	International Organization for Standardization
ISO/TC	Int. Org. for Standardization / Technical Committee
ISO/DP	Int. Org. for Standardization / Draft Proposal
DIN	Deutsches Institut für Normung e.V.
FAKRA	Normenausschuß Kraftfahrzeuge im DIN
VDE	Verband Deutscher Elektrotechniker e.V.
FTZ	Fernmeldetechnisches Zentralamt

Chemical Symbols

$R^\bullet$	C-centered radical
$RO^\bullet$	Alkoxy radical

ROO$^\bullet$	Peroxy radical
R(O)OO$^\bullet$	Acylperoxy radical
$^\bullet$OH	Hydroxyl radical
R^i, R$'$	Alkyl group
Ar	Phenyl group
CB	Carbon black

Physical Symbols

M_w	Weight-average molecular weight
M_n	Number-average molecular weight
M_w/M_n	Polydispersity
PDI	Polydispersity index
M_c	Critical molar mass
#	Number
η_0	Melt viscosity / low shear viscosity
ε	Extinction coefficient
T	Temperature
Pa	Pascal
t	Time
s	Second
MFR	Melt mass-flow rate
MVR	Melt volume-flow rate
RA	Mean surface roughness
Y.I.	Yellowness index
T_g	Glass transition temperature
T_m	Melting temperature
T_c	Ceiling temperature
C	Concentration
D	Diffusion coefficient
S	Solubility
P	Permeability
r	Reaction rate
Δ	Symbol for Heat
τ	Shear stress
Φ^{-1}	Thickness of oxidized layer

Appendix 2

Melt mass-flow rate [MFR]	ISO 1133
Melt volume-flow rate [MVR]	ISO 1133
Solution viscosity	ISO 1628
Loss modulus	ISO 537
Storage modulus	ISO 537
Elongation	ISO 527
Tensile strength	ISO 527
Tensile impact strength	ISO 8256
Charpy impact	ISO 179
Izod impact	ISO 180
Time to embrittlement	DIN 53383
Color [Y.I.]	ASTM D 1925-77
Haze	ASTM D 1003
Gloss	ASTM D 523
Hommel surface roughness	ISO/DIS 4287/1; DIN 4762/1E
Oxygen induction time [O.I.T.]	ASTM D 3895
Carbonyl index	DIN 53383
Natural weathering	ISO 4607
Artifical weathering	ISO 4892
Dry	ISO 4892 part 1
Dry / wet cycle	ISO 4892 part 1
Xenon arc	ISO 4892 part 2
Fluorescent sunlamp	ISO 4892 part 3
Carbon arc	ISO 4892 part 4

Appendix 3

Phenolic AO's

Code	Structure	CAS Reg.Nr.	Trade Name	Producer(s)
AO-1	Monophenols	10191-41-0	• Ronotec 201 • α-Tocopherol • Various others	• Hoffmann–La Roche • Various others
AO-2		128-37-0	• Ionol • Lowinox BHT • Naugard BHT • Dalpac 4 • Topanol OC • Various others	• Shell • Great Lakes • Uniroyal • Hercules • ICI • Various others

Phenolic (cont.)

Code	Structure	CAS Reg.Nr.	Trade Name	Producer(s)
AO-3	(structure: $(CH_2)_2-COC_{18}H_{37}$ ester of 3,5-di-tert-butyl-4-hydroxyphenyl)	2082-79-3	• Anox PP18 • Irganox 1076 • Lowinox PO35 • Naugard 76 • Various others	• Great Lakes • Ciba Specialty Chemicals • Great Lakes • Uniroyal • Various others
AO-4	(structure: $(CH_2)_2-C-O-C_8H_{17}$ ester of 3,5-di-tert-butyl-4-hydroxyphenyl)	12643-61-0	• Irganox 1135	• Ciba Specialty Chemicals
AO-5	Bis-phenols (structure: methylene-bridged bis-phenol)	119-47-1	• Cyanox 2246 • Irganox 2246 • Lowinox 22M46 • Oxi-Chek 114 • Vanox 2246 • Various others	• Cytec • Ciba Specialty Chemicals • Great Lakes • Ferro • Vanderbilt • Various others

Phenolic (cont.)

Code	Structure	CAS Reg.Nr.	Trade Name	Producer(s)
AO-6		35074-77-2	• Irganox 259	• Ciba Specialty Chemicals
AO-7		23128-74-7	• Irganox 1098	• Ciba Specialty Chemicals
AO-8		976-56-7	• Irganox 1222	• Ciba Specialty Chemicals
AO-9		65140-91-2	• Irganox 1425	• Ciba Specialty Chemicals

Phenolic (cont.)

Code	Structure	CAS Reg.Nr.	Trade Name	Producer(s)
AO-10		36443-68-2	• Irganox 245	• Ciba Specialty Chemicals
AO-11		85-60-9	• Santowhite Powder	• Monsanto
AO-12		90498-90-1	• Sumilizers GA 80	• Sumitomo
AO-13		1709-70-2	• Alvinox 100 • Ethanox 330 • Irganox 1330 • Various others	• Sigma • Ethyl • Ciba Specialty Chemicals • Various others

Phenolic (cont.)

Code	Structure	CAS Reg.Nr.	Trade Name	Producer(s)
AO-14		1843-03-4	• Topanol CA	• ICI
AO-15		34137-09-2	• Goodrite 3125 • Irganox 3125 • Vanox SKT	• B.F. Goodrich • Ciba Specialty Chemicals • Vanderbilt
AO-16		27676-62-6	• Goodrite 3114 • Irganox 3114	• B.F. Goodrich • Ciba Specialty Chemicals

Phenolic (cont.)

Code	Structure	CAS Reg.Nr.	Trade Name	Producer(s)
AO-17		40601-76-1	• Cyanox 1790 • Irganox 1790	• Cytec • Ciba Specialty Chemicals
AO-18	Tetra-phenols	6683-19-8	• Irganox 1010 • Anox 20 • Adekastab AO-60 • Various others	• Ciba Specialty Chemicals • Great Lakes • Asahi Denka • Various others
AO-19		32509-66-3	• Hostanox 03	• Hoechst

Phenolic (cont.)

Code	Structure	CAS Reg.Nr.	Trade Name	Producer(s)
AO-20		31851-03-3	• *Wingstay L* • *Various others*	• *Goodyear* • *Various others*

Phenolic AO's with Dual Functionality

Code	Structure	CAS Reg.Nr.	Trade Name	Producer(s)
AO-21		96-69-5	• Irganox 415 • Lowinox 44S36 • Santonox R • Santowhite Crystals • Various others	• Ciba Specialty Chemicals • Great Lakes • Monsanto • Monsanto • Various others
AO-22		90-66-4	• Irganox 1081	• Ciba Specialty Chemicals
AO-23		110553-27-0	• Irganox 1520	• Ciba Specialty Chemicals

Phenolic AO's with Dual Functionality

Code	Structure	CAS Reg.Nr.	Trade Name	Producer(s)
AO-24		41484-35-9	• Irganox 1035	• Ciba Specialty Chemicals
AO-25		991-84-4	• Irganox 565	• Ciba Specialty Chemicals
AO-26		103-99-1	• Suconox 18	• Miles
AO-27		63843-89-0	• Tinuvin 144	• Ciba Specialty Chemicals

Phenolic AO's with Dual Functionality

Code	Structure	CAS Reg.Nr.	Trade Name	Producer(s)
AO-28		4221-80-1	• Tinuvin 120 • UV-Chek AM-340	• Ciba Specialty Chemicals • Ferro
AO-29		67845-93-6	• Cyasorb UV 2908	• Cytec
AO-30		61167-58-6	• Irganox 3052 • Sumilizer GM	• Ciba Specialty Chemicals • Sumitomo

Phenolic AO's with Dual Functionality

Code	Structure	CAS Reg.Nr.	Trade Name	Producer(s)
AO-31		128961-68-2	• Sumilizer GS	• Sumitomo

Aminic AO's

Code	Structure	CAS Reg.Nr.	Trade Name	Producer(s)
AO-32		135-88-6	• Vulkanox PBN	• Bayer
AO-33		26780-96-1	• Agerite Resin D • Flectol H	• Vanderbilt • Monsanto
AO-34		101-72-4	• Vulkanox 4010 NA • Various others	• Bayer • Various others
AO-35		90-30-2	• Vulkanox PAN • Nonox AN	• Bayer • ICI
AO-36		68411-46-1	• Irganox 5057	• Ciba Specialty Chemicals
AO-37		10081-67-1	• Naugard 445	• Uniroyal

Metal Desactivators

Code	Structure	CAS Reg.Nr.	Trade Name	Producer(s)
MD-1		32687-78-8	• Irganox MD-1024	• Ciba Specialty Chemicals
MD-2		70331-94-1	• Naugard XL-1	• Uniroyal
MD-3		6629-10-3	• Eastman Inhibitor OABH	• Eastman

" Q U E N C H E R S "

Code	Structure	CAS Reg.Nr.	Trade Name	Producer(s)
Q-1		14516-71-3	• Chimassorb N-705 • Cyasorb UV-1084	• Ciba Specialty Chemicals • Cytec
Q-2		20649-88-1	• UV-Chek AM 101	• Ferro
Q-3		56557-00-7	• Sanduvor NPU	• Clariant

Phosphites / Phosphonites

Code	Structure	CAS Reg.Nr.	Trade Name	Producer(s)
PS-1		26523-78-4	• Irgafos TNPP • Various others	• Ciba Specialty Chemicals • Various others
PS-2		31570-04-4	• Irgafos 168	• Ciba Specialty Chemicals
PS-3		26741-53-7	• Ultranox 626	• General Electric
PS-4		80693-00-1	• Mark PEP-36	• Asahi Denka

Phosphites / Phosphonites

Code	Structure	CAS Reg.Nr.	Trade Name	Producer(s)
PS-5		140221-14-3	• Mark HP-10	• Asahi Denka
PS-6		38613-77-3	• Sandostab P-EPQ • Irgafos P-EPQ	• Clariant • Ciba Specialty Chemicals
PS-7		118337-09-0	• Ethanox 398	• Ethyl

Phosphites / Phosphonites

Code	Structure	CAS Reg.Nr.	Trade Name	Producer(s)
PS-8		3806-34-6	• Weston 618	• General Electric
PS-9		80410-33-9	• Irgafos 12	• Ciba Specialty Chemicals

Thiosynergists

Code	Structure	CAS Reg.Nr.	Trade Name	Producer(s)
TS-1	$\left[\ H_{37}C_{18}O-\overset{\overset{\displaystyle O}{\|\|}}{C}-CH_2-CH_2-S\ \right]_2$	693-36-7	• Argus DSTDP • Cyanox STDP • Evanstab 18 • Irganox PS 802 • Lowinox DSTDP • Various others	• Asahi Denka • Cytec • Evans • Ciba Specialty Chemicals • Great Lakes • Various others
TS-2	$\left[\ H_{25}C_{12}O-\overset{\overset{\displaystyle O}{\|\|}}{C}-CH_2-CH_2-S\ \right]_2$	123-28-4	• Argus DLTDP • Cyanox LTDP • Evanstab 12 • Irganox PS 800 • Lowinox DLTDP • Various others	• Asahi Denka • Cytec • Evans • Ciba Specialty Chemicals • Great Lakes • Various others

Thiosynergists

Code	Structure	CAS Reg.Nr.	Trade Name	Producer(s)
TS-3	$\left[H_{29}C_{14}O-\overset{\overset{\displaystyle O}{\|}}{C}-CH_2-CH_2-S \right]_2$	16545-54-3	• Argus DMTDP • Cyanox MTDP • Evanstab 14 • Various others	• Asahi Denka • Cytec • Evans • Various others
TS-4	$\left[H_{37}C_{18}-S \right]_2$	2500-88-1	• Hostanox SE-10	• Hoechst

UV-Absorbers

Code	Structure	CAS Reg.Nr.	Trade Name	Producer(s)
UVA-1	Benzophenones — benzophenone with OH and OCH₃ substituents (OCH_3)	131-57-7	• Cyasorb UV-9 • Syntase 62	• Cytec • Great Lakes
UVA-2	benzophenone with OH and OC_8H_{17} substituents	1843-05-6	• Chimassorb 81 • Sumisorb 130 • Cyasorb UV-531 • Syntase	• Ciba Specialty Chemicals • Sumitomo • Cytec • Great Lakes
UVA-3	benzophenone with OH and $OC_{12}H_{25}$ substituents	2985-59-3	• Eastman DOBP • Syntase 1200	• Eastman • Great Lakes

UV-Absorbers

Code	Structure	CAS Reg.Nr.	Trade Name	Producer(s)
UVA-4		43221-33-6	• Seesorb 1000	• Shipro Kasei
UVA-5		57472-50-1	• Mark LA-51	• Asahi Denka
UVA-6	Benzotriazoles	2440-22-4	• Mark LA-32 • Tinuvin P • Various others	• Asahi Denka • Ciba Specialty Chemicals • Various others

UV-Absorbers

Code	Structure	CAS Reg.Nr.	Trade Name	Producer(s)
UVA-7		3147-75-9	• Cyasorb UV-5411 • Tinuvin 329 • Various others	• Cytec • Ciba Specialty Chemicals • Various others
UVA-8		3896-11-5	• Mark LA-36 • Tinuvin 326 • Various others	• Asahi Denka • Ciba Specialty Chemicals • Various others
UVA-9		3846-71-7	• Tinuvin 320 • Various others	• Ciba Specialty Chemicals • Various others
UVA-10		23328-53-2	• Tinuvin 571	• Ciba Specialty Chemicals

UV-Absorbers

Code	Structure	CAS Reg.Nr.	Trade Name	Producer(s)
UVA-11		25973-55-1	• Tinuvin 328 • Various others	• Ciba Specialty Chemicals • Various others
UVA-12		36437-37-3	• Tinuvin 350	• Ciba Specialty Chemicals
UVA-13		3864-99-1	• Mark LA-34 • Tinuvin 327 • Various others	• Asahi Denka • Ciba Specialty Chemicals • Various others
UVA-14		70321-86-7	• Tinuvin 234	• Ciba Specialty Chemicals

UV-Absorbers

Code	Structure	CAS Reg.Nr.	Trade Name	Producer(s)
UVA-15		103 597-45-1	• Mark LA-31 • Tinuvin 360 • Various others	• Asahi Denka • Ciba Specialty Chemicals • Various others
UVA-16		84268-08-6	• Tinuvin 840	• Ciba Specialty Chemicals

UV-Absorbers

Code	Structure	CAS Reg.Nr.	Trade Name	Producer(s)
UVA-17	Hydroxyphenyl-triazines	147315-50-2	• Tinuvin 1577	• Ciba Specialty Chemicals
UVA-18		2725-22-6	• Cyasorb UV-1164	• Cytec

UV-Absorbers

Code	Structure	CAS Reg.Nr.	Trade Name	Producer(s)
UVA-19	Oxanilides	23949-66-8	• Tinuvin 312 • Sanduvor VSU	• Ciba Specialty Chemicals • Clariant
UVA-20		35001-52-6	• Tinuvin 315 • Sanduvor EPU	• Ciba Specialty Chemicals • Clariant
UVA-21	Cinnamates	7443-25-6	• Cyasorb UV-1988	• Cytec

Hindered Amines "HAS"

Code	Structure	CAS Reg.Nr.	Trade Name	Producer(s)
HAS-1	Low Molecular Weight HAS $O-C-(CH_2)_n-CH_3$ n = 14,16,18-Mixture	24860-22-8	• Cyasorb UV-3853-S	• Cytec
HAS-2	$C_{12}H_{25}$ ... (structure)	79720-19-7	• Cyasorb UV-3581	• Cytec
HAS-3	$C_{12}H_{25}$ ... (structure)	106917-30-0	• Cyasorb UV-36041	• Cytec
HAS-4	(structure) $(CH_2)_9$	64338-16-5	• Hostavin N-20	• Hoechst

Hindered Amines "HAS"

Code	Structure	CAS Reg.Nr.	Trade Name	Producer(s)
HAS-5		52829-07-9	• Tinuvin 770 • Sanol LS-770	• Ciba Specialty Chemicals • Sankyo
HAS-6		41556-26-7	• Tinuvin 765	• Ciba Specialty Chemicals
HAS-7		99473-08-2	• Sumisorb LS-060	• Sumitomo
HAS-8		71029-16-8	• Goodrite UV-3034	• B.F.Goodrich

Hindered Amines "HAS"

Code	Structure	CAS Reg.Nr.	Trade Name	Producer(s)
HAS-9	High Molecular Weight HAS	65447-77-0	• *Tinuvin 622*	• *Ciba Specialty Chemicals*
HAS-10		71878-19-8	• *Chimassorb 944*	• *Ciba Specialty Chemicals*
HAS-11		90751-07-8	• *Cyasorb UV-3346*	• *Cytec*

Hindered Amines "HAS"

Code	Structure	CAS Reg.Nr.	Trade Name	Producer(s)
HAS-12		78276-66-1	• Hostavin N-30	• Hoechst
HAS-13		154636-38-1	• Uvasorb HA 88	• Sigma

Hindered Amines "HAS"

Code	Structure	CAS Reg.Nr.	Trade Name	Producer(s)
HAS-14		100631-44-5	• Mark LA-68	• Asahi Denka
HAS-15		115055-30-6	• Mark LA-63	• Asahi Denka
HAS-16		115810-23-6	• Chimassorb 119	• Ciba Specialty Chemicals

Hindered Amines "HAS"

Code	Structure	CAS Reg.Nr.	Trade Name	Producer(s)
HAS-17		164648-93-5	• Uvasil 299	• Great Lakes
HAS-18		96204-36-3	• Goodrite 3150	• B.F.Goodrich

Hindered Amines "HAS"

Code	Structure	CAS Reg.Nr.	Trade Name	Producer(s)
HAS-19		130277-45-1	• Goodrite 3159	• B.F. Goodrich

Subject Index

M. Braden, R.L. Clarke, London, UK;
J. Nicholson, Teddington, UK; **S. Parker,** London, UK

Polymeric Dental Materials

1997. XII, 124 pp. 32 figs., some in color, 34 tabs.
(Macromolecular Systems - Materials Approach)
Hardcover DM 198
ISBN 3-540-61646-2

Braden and his co-authors give a comprehensive
overview of the use of polymers and polymer
composites as dental materials. These comprise
polyelectrolyte based materials, elastomers, glassy
and crystalline polymers and fibres. Such materials
are used in dentistry as restorative materials, hard
and soft prostheses, and impression materials.
The chemistry of materials is reviewed, together
with mechanical, thermal, visco-elastic and water
solution properties. These properties are related to
clinical performance, with emphasis on some of the
difficulties inherent in developing materials for
oral use. Indications are given of possible future
developments.

■ ■ ■ ■ ■ ■ ■ ■ ■ ■

Please order from
Springer-Verlag Berlin
Fax: + 49 / 30 / 8 27 87- 301
e-mail: orders@springer.de
or through your bookseller

Price subject to change without notice.
In EU countries the local VAT is effective.

Springer

Springer-Verlag, P. O. Box 31 13 40, D-10643 Berlin, Germany

V. Shibaev, Moscow State University,
Moscow, Russia (Ed.)

Polymers as Electrooptical and Photooptical Active Media

1996. XVII, 211 pp. 117 figs., 30 tabs.
(Macromolecular Systems - Materials Approach)
Hardcover DM 148
ISBN 3-540-59486-8

Contents: Photochemical Hole Burning and
Photooptical Properties of Doped Dye Molecules
in Linear Polymers. – Comb-Shapoed Polymers
with Mesogenic Side Groups as Electro- and
Photooptical Active Media. – Cyclic Liquid
Crystalline Siloxanes as Optical Recording
Materials. – Photoregulation of Liquid Crystal
Alignment by Photochromic Molecules and
Polymeric Thin Films. – Electric Field Poling of
Nonlinear Optical Side Chain Polymers.

Please order from
Springer-Verlag Berlin
Fax: + 49 / 30 / 8 27 87- 301
e-mail: orders@springer.de
or through your bookseller

Price subject to change without notice.
In EU countries the local VAT is effective.

Springer

Springer-Verlag, P. O. Box 31 13 40, D-10643 Berlin, Germany

H. Pasch, Darmstadt, Germany;
B. Trathnigg, Universität Graz, Austria

HPLC of Polymers

1997. Approx. 220 pp. 122 figs., 29 tabs.
(Springer Laboratory)
Hardcover DM 128
ISBN 3-540-61689-6

This lab manual describes the different
chromatographic methods for polymers, their
fundamentals, equipment, experimental
procedures and applications. As a result, the
book enables polymer chemists, physicists and
material scientists, as well as students of
macromolecular and analytical science to
optimize chromatographic conditions for a
specific separation problem.

Please order from
Springer-Verlag Berlin
Fax: + 49 / 30 / 8 27 87- 301
e-mail: orders@springer.de
or through your bookseller

Price subject to change without notice.
In EU countries the local VAT is effective.

Springer

Springer-Verlag, P. O. Box 31 13 40, D-10643 Berlin, Germany

Springer
and the
environment

At Springer we firmly believe that an international science publisher has a special obligation to the environment, and our corporate policies consistently reflect this conviction.

We also expect our business partners – paper mills, printers, packaging manufacturers, etc. – to commit themselves to using materials and production processes that do not harm the environment. The paper in this book is made from low- or no-chlorine pulp and is acid free, in conformance with international standards for paper permanency.

Printing: Saladruck, Berlin
Binding: Buchbinderei Lüderitz & Bauer, Berlin